I0820496

THE RUSSIAN SPACE PROGRAM

This book is dedicated to the memory of those cosmonauts and astronauts who gave their lives in the pursuit of space exploration.

'If we die, we want people to accept it. We are in a risky business, and we hope that if anything happens to us, it will not delay the program. The conquest of space is worth the risk of life.'

Lieutenant Colonel Virgil Ivan 'Gus' Grissom – Astronaut
Mercury 7 (*Liberty Bell*), *Gemini 3* and *Apollo 1*.

B KOCMOC, INTO THE COSMOS

TERRY C. TREADWELL

AN IMPRINT OF PEN & SWORD BOOKS LTD.
YORKSHIRE – PHILADELPHIA

First published in Great Britain in 2025 by
White Owl
An imprint of
Pen & Sword Books Ltd.
Yorkshire – Philadelphia

ISBN 978 1 03611 727 6

A CIP catalogue record for this book is available from the British Library.

Typeset in 11/13 pts Cormorant Infant
by SJmagic DESIGN SERVICES, India.

The Publisher's authorised representative in the EU for product safety is Authorised Rep Compliance Ltd., Ground Floor, 71 Lower Baggot Street, Dublin D02 P593, Ireland.
www.arccompliance.com

For a complete list of Pen & Sword titles please contact

PEN & SWORD BOOKS LIMITED
George House, Beevor Street, Off Pontefract Road, Hoyle Mill, Barnsley,
South Yorkshire, England, S71 1HN.
E-mail: enquiries@pen-and-sword.co.uk
Website: www.pen-and-sword.co.uk

or

PEN AND SWORD BOOKS
1950 Lawrence Rd, Havertown, PA 19083, USA
E-mail: uspen-and-sword@casematepublishers.com
Website: www.penandswordbooks.com

Contents

Introduction

Although the success of the high profile American space program had grabbed all the headlines, Russia slowly, but consistently, has continued to expand their contribution to the exploration of space. The survival of the International Space Station (ISS) is now almost the prime combined responsibility of Russia and America, and since the phasing out of the Space Shuttle, it is the Russians that now have the main responsibility to replace the crew members and take supplies to the space station, although now the American SpaceX program has eased that burden.

In the formative years of space exploration, it was Russia who led the way when they placed a satellite in orbit and then put the first man, Yuri Gagarin, into space. They quickly followed this with a succession of manned and unmanned missions and, unfortunately, a few disasters and casualties along the way. Because of the secretive nature of Russia at the time, which remains to a certain degree, these casualties never came to light until many years later, when the understanding and tensions between the East and West eased.

Despite the number of setbacks that dogged the early years, Russia slowly developed spacecraft that, although considered by some to be very basic, were reliable. So, from a very slow and shaky start, the Russian space program progressed into being one of the foremost leaders in the aim to explore space, and has brought two of the most powerful nations on Earth closer together. Although the development of the *Salyut* and *Mir* space stations created a closer understanding between the nations, the work and experiments carried out by the cosmonauts in these early space stations were almost all military and were either secret or restricted, so little is known about them. However with the arrival of the International Space Station came a degree of openness, and great strides were made in the understanding of the physiological and psychological effects on the men and women involved in working in space and a weightless environment.

Unfortunately, since the war in the Ukraine, Russia's relationship with the West has deteriorated considerably and co-operation within the space program is very limited, although the two main powers have reached an agreement regarding the transportation of crews and supplies to the ISS. It is thought, however, that the International Space Station's days are numbered and it will close in 2031.

This book also makes reference to only a small number of Kosmos missions; those mentioned are the only ones with any connection to space exploration. The remainder, and there have been over 2,500 of them, were either military or scientific satellites.

CHAPTER ONE

The Birth of the Rocket in Russia

Although the history of rockets goes back many, many, centuries, in Russia no records existed of their use until the 1600s, although it is almost certain that rockets were in existence in the country's arsenals. Documents written between 1607 and 1621 were, in 1777, accumulated by the Russian gunsmith Onisin Mikhailov and turned into a compiled document entitled '*Code of Military, Artillery and Other Matters Pertaining to the Science of Warfare*'. In this document there were detailed descriptions of rockets, referred to as 'Cannon balls which run and burn'. The reliability of this early document was disputed by many, mainly because the manuscript contained information which was not entirely Mikhailov's. It was in fact a collection of some 663 snatches of information and articles from a variety of foreign military books and sources. It doesn't really matter that Mikhailov collected and published these articles, calling them his own work, what is of more importance is the fact that he published them and by doing so opened up a new world.

Tsar Peter the Great of Russia (1672–1725) devoted a lifetime to creating Russia's military might, and in 1680 he founded the first Rocket Works in Moscow. There they made illuminating and signal rockets for the army, all used under the guidance of English, Dutch, German and French officers, who instructed the army in their use. Then, in the early 1700s, the Tsar made St. Petersburg the new capital of Russia, and moved the entire Rocket Works to this new location, expanding it at the same time.

The development of Russia's rockets continued into the late 1700s, when a senior officer in the Tsar's artillery, General Alexander Dimitrievich Zasyadko (1774–1837), who had been studying Sir William Congreve's (1772–1828) progress and exploits in the use of rockets, together with the files from the Rocket Works, decided to design some rockets of his own. After taking part in a number of wars, including the Italian War in 1799, the Russo–Turkish War (1806–1812) and the Napoleonic War of 1812, he concentrated on creating military gunpowder and weapons. Amongst his creations were rocket-launching platforms capable of firing six rockets in salvos and a number of gun-laying devices.

So successful were the tests of his rockets, that in 1817 Zasyadko was assigned to western Russia to train the Tsar's soldiers in the use of military rockets. The following year a school of artillery was opened and Zasyadko was appointed its

head with a promotion to Major General, he was also appointed head of the St. Petersburg Armoury and the Okhtensky Gunpowder Factory. It was during this period that he took part in the second Russo–Turkish War (1828–1829), where his solid fuel rockets were used during the sieges of Varna, Braila, Silistra and Schmia, earning him a promotion to Lieutenant General. On the Black Sea, Russian ships used the rockets with great success against the Turkish fleet and on the shore-based gun emplacements. During the insurrection in the Caucasus, where a large number of small battles took place, thousands of rockets were used with increasing devastation and greater reliability. They were also used later in the Crimea War and again with positive results.

The death of Lieutenant General Alexander Zasyadko in 1837 caused his position as head of the school to be taken by another artillery officer by the name of Konstantin I. Konstantinov. This 30-year-old officer was the first to work on the practical problems of rocket production and became the founder of experimental rocket dynamics. Up to this point, the production of rockets had been left to the individual skill of the makers and, as can be imagined, there were a number of accidents. Among Konstantinov's achievements was the development of large-scale rocket production, bringing together almost all of the skilled engineers. Konstantinov also looked at non-military applications, one of which resulted in the development of a rocket that could fire lifelines to wrecked vessels. This device was to be responsible for saving hundreds of lives in the forthcoming years and even today the principle is still used, albeit with a more powerful and sophisticated model.

Towards the end of the nineteenth century interest in military rockets started to decline, mainly because of the increased development and reliability of long-range artillery. But as one door starts to closes, another one opens and interest in the use of rockets as a propellant in aviation increased. One Russian artillery officer by the name of Iustin Ivanovich Tretesky submitted a design for a dirigible involving the use of rockets, but it was rejected as being completely impractical. He later submitted two more designs for a dirigible powered by gunpowder and compressed air and once

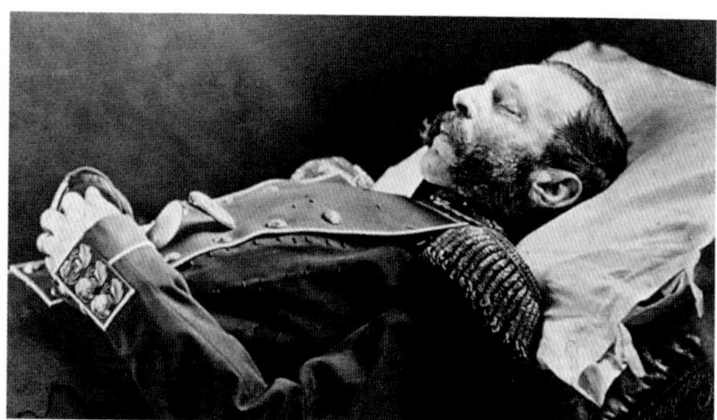

Above left: Nikolai Kilbachich. (Academy of Sciences)

Above right: Tsar Alexander II lying in state after being assassinated by a bomb made by Nikolai Kilbachich. (Author)

again both were rejected as unworkable. However his idea was endorsed in 1881, when Nikolai I. Kilbalchich proposed the idea of heavier-than-air machines with rocket propulsion and carrying human passengers. Unfortunately Kibalchich also used his enthusiasm and expertise in another direction. He was the bomb expert and bomb maker for the revolutionary organization *Naradonaya Volya* (People's Will), and a bomb thrown at Tsar Alexander II, which mortally wounded him, was found to have been prepared by Kibalchich. He was arrested by the Tsar's secret police and executed soon afterwards.

Around this period, there was never a shortage of inventions and designs for airships and rockets, most of which were either impracticable or pure flights of fancy, but amongst them there were the odd one or two that had merit. One of these was by an engineer by the name of A.V. Evald, who in 1886 submitted a proposal for an aircraft powered by a type of engine that had rockets packed tightly in a specially designed tube as its power source. There were a number of unsuccessful tests carried out and the project was abandoned after the money ran out, despite the fact that the last test was completely successful. Another inventor was Feodor Geshvend, who in 1887 put forward a design for a steam-jet powered aircraft which would have the capability of hovering in the air. The aircraft got no further than the design stage as it was considered totally impractical.

A number of other Russian inventors produced ideas over the next few years, but modern rocketry started with the innovations of Konstantin Eduardovich Tsiolkovsky (1857–1935), considered by many Russians to be the 'Father of Soviet Space Flight'. Born in 1857, the son of a forestry expert and inventor, Tsiolkovsky enjoyed a normal childhood. Then, when he was eight years old, he contracted scarlet fever and became almost totally deaf. Unable to go to school, he taught himself from his father's books, first mastering mathematics and then physics. Such was his progress that his father sent him to the Technical School in Moscow to further his education. After three years at the Technical School he returned to his hometown and, despite his lack of formal education and qualifications, was appointed as a teacher. It was whilst he was teaching that he started to carry out serious research in the areas of the airplane, an all-metal dirigible, and a rocket for interplanetary travel. The development of the rocket was the one idea that attracted his attention more than the others and he concentrated his efforts in that direction. However he still retained an interest in the airplane and the dirigible, and produced designs for both.

Konstantin Eduardovich Tsiolkovsky. (Gromov Institute)

In 1885, Tsiolkovsky had produced a design for an all-metal airship that had

movable sides, enabling the volume of gas to be varied so that the craft could maintain constant lift at alternating air temperatures and altitudes. By placing the gas bags close to the engine's exhaust, the gas could be heated and expanded. The proposal was rejected by both the Russian Engineering Society and the General Staff of the Imperial Russian Army. Despite the rejections there were a number of fellow enthusiasts and engineers who supported him.

Konstantin Tsiolkovsky was a man of vision and was years ahead of his time in his concepts and designs. Between 1885 and 1895, Tsiolkovsky carried out a great deal of work on the design of metal airships and at first was considered to be an eccentric, living in a world of fantasy. In 1894 he proposed a design for an all-metal aeroplane in an article entitled, '*The Airplane, A Bird-like Flying Machine*', in which he gave a detailed description, supported by a series of sketches, of a monoplane with rounded wingtips and a streamlined fuselage. Some twenty years later, very similar aircraft designs appeared in other European countries and were hailed as being far ahead of their time.

With the lack of success with dirigibles, Tsiolkovsky turned his attention to the development of aerodynamics. In 1890, he realised that as aerodynamics progressed there was going to be a need for more sophisticated test equipment and so he built a wind tunnel enabling him to carry out hundreds of tests. This led him into the world of using rockets for space exploration and in 1903 he published his article, '*Investigating Space with Reaction Devices*'. In it he described his formula that showed it was possible to determine the flight performance of a rocket, if the weights of the propellant were known, enabling it to reach orbital height and have escape velocity. But he was too far ahead of his time and in the main he was mostly ignored and the paper went unnoticed. After years of research, design and lecturing, Konstantin Tsiolkovsky was finally recognized and elected into the Russian Socialist Academy. With the election into the academy came a personal pension from the 'Commission for Improvement of the lot of Scientists' (TsEKUBU). The Socialist Academy was later named the Communist Academy and later still The Academy of Science, USSR.

Friedrich Tsander. (Academy of Sciences)

A Lithuanian engineer by the name of Friedrich Tsander (1887–1933), an ardent follower of Tsiolkovsky, came to prominence in 1908 when, after corresponding with the scientist, he was asked to edit a compilation of the great man's work. Over the following years Tsander, who had become a close friend of Tsiolkovsky, designed a number of rocket planes, engines and boosters, and developed a reputation for lecturing on the ideas of space travel. So dedicated to the science of space travel and all that it encompassed, when he married, he named his daughter and son Astra and Mercury.

Tsander's expertise covered three areas; the development of the rocket engine, the building and testing of rocket engines and the problems that could be associated with spaceflight. The latter, of course, covered the problem of escaping from the Earth's gravitational field. The development of rockets to be used in space travel was the area in which Tsander showed the greatest interest. He favored using a multi-stage rocket incorporated into an airplane. The idea was that the airplane would fly the rocket to a high altitude at which point the rocket would then be ignited to push itself into space. In 1924, together with Konstantin Tsiolkovsky, he co-founded the Society for the Studies of Interplanetary Travel. Between them they published articles on human spaceflight and later helped in the design of Russia's first liquid-fueled rocket.

In 1928, Tsander developed the OR.1 (Opytnyi Reaktivny) engine using compressed air and gasoline as the propellants. Later that year the engine was put into a glider, Opel-RAK, and with Fritz von Opel at the controls, carried out the first rocket-powered flight. With the success of this engine, which completed more than 50 tests, work began on developing the OR.2 engine powered by benzene and liquid oxygen. This engine was successfully tested in March 1933, and plans were drawn up to put it into an experimental glider known as the R-1.

Then tragedy struck. Just ten days after the successful testing of the OR.2 engine, Friedrich Tsander contracted typhoid fever and died. A memorial to him and his

First flight of the OPELRAK 1 rocket-propelled glider. (Author)

Fist flight of OPELRAK 2 rocket-propelled glider. (Author)

work was later erected in the town of Kislovodsk. His work continued with his ideas and designs being made available to the science academies, which allowed interested engineers to use them. Amongst these was the man known as Yuri Kondratyuk, an engineer from Poltava, Ukraine. Born Oleksandr Ignatyevich Shargei on 21 June 1897, his mother, Ludmila Lvovna Schlippenbach, was said to have been a descendant of the Swedish General Wolmar Anton von Schlippenbach, who, in 1709, led the Swedish army's failed invasion of Russia in the time of Charles XII of Sweden. Yuri Kondratyuk's father, Ignatiy Benediktovich Shargei, studied mathematics and physics at Kiev University and became a lecturer. Both Yuri Kondratyuk's parents were intellectuals, but his mother, having been arrested on a number of occasions for demonstrating on social reform, was deemed by the authorities to be suffering from a mental disorder and was incarcerated. Yuri's father divorced her and it was left to his paternal grandmother to bring him up.

Yuri Kondratyuk. (Roscosmos)

On leaving school Oleksandr Shargei went to the Peter the Great Polytechnic in Petrograd where he studied engineering. At the onset of the First World War he joined the army as a junior officer, but all the time he was fighting at the front his mind was elsewhere. By the end of the war he had filled four notebooks with drawings of spacecraft and calculations on how to send them into space.

The end of the First World War saw the start of the Russian Revolution and an increase of great poverty in the country. Surviving members of the Tsarist's Army were considered to be enemies of the state and were hunted down. Oleksandr contracted typhus and was nursed back to health by friends who also found him a new identity – Yuri Vasilievich Kondratyuk. He went to work in Siberia as a mechanic and whilst there, completed his book '*The Conquest of Interplanetary Space*'. As no publisher would publish 'this flight of fantasy' as one put it, Yuri published the book himself. Other engineers enthusiastically read it and 2,000 copies were printed. His engineering expertise came to the fore in the early 1920s, when he designed and built a massive grain elevator that was constructed without the use of a single nail. The success of the grain elevator was admired by almost everyone until, during one of Joseph Stalin's purges in the 1930s, the NKVD (Naródnyĭ Komissariát Vnútrennikh Del) arrested him for being a saboteur. They used the fact that the grain elevator had been built without nails as an excuse to accuse Kondratyuk of planning the structure to collapse. The fact that the structure had been there for some years without any problems seemed to have eluded them. After being sentenced to three years in a labour camp, his engineering expertise was quickly recognized and he was sent to the Kubass coal-mining region to evaluate foreign machinery. However, his reputation for the building of the grain elevator had preceded him and he was transferred almost immediately to Siberia to work on grain projects. It was while here that he became involved with the design of a wind power generator scheme for the Crimea. His design for a 500ft tower, to which was fitted a propeller with a 240ft span, was submitted and accepted. He was sent to Moscow to meet with Sergo Ordzhonikidze, the Commissar for Heavy Industry.

Sergei Korolev inspecting German rockets at Peenemunde. (Author)

It was also on this visit to Moscow that he met Sergei Korolev, the head of the rocket research group Gruppa Izucheniya Reaktivnogo Dvizheniya (GIRD), who, having read Kondratyuk's book and other works of his, offered him a job. Tempting though it was, Kondratyuk refused, worried that his true identity would be revealed during any security check. Despite this he continued with his research into the launching and recovery of spacecraft. It is uncanny that his ideas of landing on the

Moon and returning back to Earth are very similar to those followed by the Apollo missions. His theory for re-entry was to approach the Earth without reducing speed, and use the atmosphere to not only slow the speed of the spacecraft down, but also to capture the spacecraft using the Earth's gravity.

At the outbreak of the Second World War Kondratyuk joined the Russian Army and fought against the Germans, but disappeared during the fighting near Zasetski. His body was never recovered. There were rumors that he went to the United States and continued his work there under an assumed identity, but there is no evidence to support this.

Tsiolkovsky's paper had been re-published in 1911 with additional information regarding propellants, but this time other scientists and engineers took notice. His work was slowly becoming accepted and he was no longer regarded as an eccentric, but it wasn't until 1918 that he was completely recognized for what he had achieved. In 1919 he was elected to the Russian Socialist Academy and granted a personal pension by the Commission for Improvement of the Lot of Scientists (TsEKUBU), later to become known as the Academy of Sciences USSR.

A number of groups and societies started to spring up around Russia, their objectives being investigation into space travel. The government supported them by setting up a department called the Central Bureau for the Study of Problems of Rockets (TsBIRP). The department's role was to bring together all the different societies and groups under one 'umbrella' so to speak, discover what strides were being made in the west, assimilate all the results and information and see if they would have any military applications. One of the new societies that sprang up was the Russian Rocket Society formed under the leadership of Professor Nikolai Rynin and Doctor Yakov Isordorovich Perelman. Both died in 1942 from starvation during the siege of Leningrad.

It was in Germany around this time that the development of high-altitude, long-range rockets started, albeit for military purposes. It is relevant to mention Germany at this point, because at the end of the war it was the capture of some of Germany's rocket scientists that was the beginning of Russia's progress into the world of space exploration. The real beginnings of rocketry, as far as the Germans were concerned, was in 1923 with the publication of a booklet called '*Die Rakete zu den Planetenaumen*' (The Rocket into Interplanetary Space) by Professor Hermann Oberth. Then in 1927 the German Interplanetary Society was formed and work began in earnest on the design and manufacture of rockets. Such was the interest, that a 300-acre aerial research facility was established just outside Berlin called the Raketenflugplatz and was placed under the control of Professor Oberth. Then in 1929 the Society for Space Travel (Verein für Raumschiffahrt) was formed. On the face of it this was a civilian operation, but it soon became obvious that the German Army had become heavily involved by supplying equipment and personnel. On testing days there were always a number of visitors in civilian clothes arriving in large black limousines, who were quite clearly senior military personnel, one of whom was Major General Walter Dornberger, later commanding officer of Peenemunde.

Major General Walter Dornberger (centre) at Peenemunde. (Author)

The Treaty of Versailles in 1919 had disarmed Germany, forbidding the manufacture of armoured cars, tanks, submarines and military aircraft, with a very limited number of factories allowed to manufacture small arms and ammunition. However, they had not placed any restrictions on the development and manufacture of rockets. This was a serious oversight by the Allies, because although the science was in its infancy, rockets had been used in warfare for centuries and there were disturbing signs that German rocket development for military purposes had started.

Meanwhile in Russia, Konstantin Tsiolkovsky, who was a great admirer of Pereleman's work, had published a book in 1929 called '*Kosmicheskiye Raketnyye Poyezda*' (Cosmic Rocket Trains) in which he described in great detail his design for a two-stage rocket. The first stage would take the rocket through the Earth's atmosphere, at which point it would separate and return to Earth, whilst the second stage would continue out into space and beyond. He then moved his attention away from rockets to the jet engine, and in 1930 wrote an article explaining the advantages and disadvantages of the jet engine. Tsiolkovsky died in 1935 a national hero, bequeathing all his papers and models to the Soviet Government. In 1952, the Aero Club of France had a large gold medal struck in his honour, and in 1954 the Soviet government established the Tsiolkovsky Gold Medal, which has been awarded every three years, since its installation, to the most outstanding contributor to space technology.

Two of the main organizations in the late 1920s were the LGDL (Leningrad Gas Dynamics Laboratory) and GIRD (Group for the Study of Reaction Propulsion).

The GDL group concentrated their efforts on designing and developing high energy, solid rocket motors. In the 1930s an engineer by the name of Valentin Petrovich Glushko carried out experimental work on rocket engines developed mainly for military purposes – aircraft rockets and later anti-tank rockets.

Then in October 1933, a story started circulating that two German rocket designer/engineering brothers, Bruno and Otto Fischer, had launched a manned rocket six miles into the air, almost to the edge of space, from Rügen, an island in the Baltic Sea. The story, printed in the British newspaper *The Sunday Referee*, was that the Reichswehr (German War Ministry) had funded the project in total secrecy. It had been rumored that a similar project had been tried the year before and had crashed, suffering a fatality. It was claimed that on the morning of 29 October 1933, Otto Fischer crawled into the rocket through a small steel door on the side, whilst his brother Bruno Fischer and three Reichswehr officials retired to a safe distance before igniting the rocket. In a blinding flash and a roar, the rocket left the frame that had been supporting it and disappeared from sight. Ten minutes later the rocket re-appeared on the end of a parachute and settled on a sandy beach nearby. A white faced, shaken Otto Fischer is said to have emerged with a big smile on his face, probably grateful that he had survived. There were a number of experts who did not believe this event ever happened: some thought it to be a German military propaganda exercise, others thought that there was possibly a semblance of truth, but there was never any comment from the Germans or any other quarter. Whatever the truth, because of the increasing unrest developing with Germany and most of Europe, it became the cause of some concern and speculation in Russia.

Not all those involved in the rocket development program were engineers; some of the most equally important ones were mathematicians and theoreticians and one of these was Mikhail Tikhonravov. He had initially started out studying the aerodynamics of birds in flight and by the 1920s had progressed to the design of rockets. He became a friend and associate of Koroltov and Glushko whilst a member of GIRD and was responsible for the design of many of the early rockets.

In March 1938, during one of Stalin's infamous purges, Valentin Glushko was arrested by the NKVD. The following August, after a brief 'trial' he was sentenced to eight years in a Gulag where he was put to work on various aviation projects together with a number of other 'arrested' scientists.

The first of the rockets from GDL was the ORM-1 (Opytnyi Reaktivnyi Motor) and had been designed by Friedrich Tsander in 1930. It was powered by gasoline and liquid oxygen with nitrogen tetroxide as the oxidizer. This was followed by the ORM-2, which was fitted with solid-propellant ignition and liquid hypergolic propellants. The latter was a propellant that ignited on contact. With the information gathered from these tests, more advanced engines were developed: the ORM-4 to ORM-22. The most outstanding engine was the ORM-52, which was chosen to power a naval torpedo, a meteorological rocket and an anti-aircraft rocket. Another of the most innovative engines Korolev designed and built, was the ORM-65. This engine was designed to be installed into both a flying bomb and a rocket-powered aircraft. But as the war

Early rocket engineers with Sergei Korolev. (Roscosmos)

progressed, the team switched their attention to the development of a new engine, the RD-1. After numerous tests the RD-1 was fitted into a twin-engined Petlyakov PE-2 fighter/bomber, a Lavochkin La-7R fighter and a Yakovlev Yak-3 fighter. The RD-1 was later superseded by the RD-2 and then the RD-3.

Sergei Korolev, who had offered Kondratyuk a job with his rocket research group, suddenly found himself on the receiving end of one of Stalin's purges in 1937. He, together with a large number of other aerospace engineers and rocket designers, was arrested and imprisoned. However, with the outbreak of the Second World War, Stalin recognized the need for aerospace engineers and set up a series of prison design bureaus, known as Sharashkas. In charge of one of these Sharashkas was the aircraft designer Andrei Tupolev and he immediately requested that Korolev be assigned to work with him.

The sudden rise of the Nazi party in Germany, coupled with their blatant disregard of the Treaty of Versailles, caused a foreboding sense of concern to sweep across Europe. Germany began to build her military forces, once again seemingly unopposed, and in 1937, the rocket research and experimental facility at Peenemunde was up and running with Major General Walter Dornberger as its commanding officer and Werner von Braun as its civilian director.

Andrei Tupolev, one of Russia's top aircraft designers. (Roscosmos)

Werner von Braun (far right) and General Walter Dornberger (centre) with other high ranking Nazi officials at Peenemunde. (Author)

Adolf Hitler granted the facility 20 million reichmarks for research purposes, but after the successful campaign in Poland and the Benelux countries, he decided that he had no need for rockets and moved his attention to Göring's Luftwaffe. Hermann Göring had convinced Hitler that his Luftwaffe would destroy the British air force, take control of the skies over Europe and would be a major factor in winning the war. The initial impetus of Hitler's military push surprised everyone and with it Hitler's support for Peenemunde and the rocket program was put to one side. Funding was withdrawn, work on various rocket projects were dramatically curtailed and the vast majority of the civilian workforce at Peenemunde conscripted into the army. Dornberger and von Braun visited Hitler asking him to reconsider his decision, but were told that Hitler had had a dream that told him the rockets would not work. There had been rumors that Hitler was an almost fanatical follower of astrology and the occult and used it to make decisions, but there is little evidence to support this.

With the collapse of the Nazi Germany/Soviet Union non-aggression pact, followed by the German attack on the Soviet Union, focus changed in Russia from space development and research to survival, and the GIRD was tasked with providing smaller rockets for the defence of the country. They produced a multi-barrelled, ground-to-ground, rocket battery later nicknamed Katyusha, that had been designed by General Andrey Kostikov, which during the violent and horrendous battle for Leningrad, was a major contributing factor in halting the German Army's advance. Throughout the war the work of GIRD was focused on armament, mainly because almost all their rocket research establishments and industrial centres were in German hands, although the majority of the staff had been moved farther East.

In 1941 Valentin Glushko was placed in charge of a design bureau for liquid-fueled rocket engines, where he met Serge Korolov. He was released in 1944 and went to work

Werner von Braun (centre) and Walter Dornberger (far left) at Peenemunde. (Author)

with Sergei Korolov, where together they designed the RD-1 KhZ auxiliary rocket motor, which was tested in a fast climbing Lavochkin La-7R fighter aircraft that had been developed for protection of the capital from high-altitude Luftwaffe bombing attacks.

Then as Hitler's push against Russia ground to a halt and Göring's Luftwaffe failed to deliver their promise to wipe the British from the skies over Europe, the German rocket development at Peenemunde suddenly took on a new lease of life and became a priority. The successful test launch of the V-2 (Vergeltungswaffe-Zwei - Vengeance Weapon No.2) rocket on 3 October 1942, followed in December by the pulse-jet powered Fieseler Fi-103 V-1 (Vergeitungswaffe Eins-Vengeance Weapon No.1), carrying a one ton warhead, attracted the attention of Hitler once again and he ordered that the V-1 flying bomb and V-2 rockets should be put into immediate mass production.

With typical German efficiency, two missile battalions were created, a mobile one and a static one. The mechanized battalion consisted of three batteries, each with three launching platforms and nine missile trailers and capable of firing twenty-seven V-2 rockets a day. The static battalion was to be based at Cherbourg and have a capability of firing thirty-six V-2 rockets per day. There were to be three assembly plants, one in the east, which was to be located close to the city of Riga in Latvia, one in the centre to be located at the southern end of the Harz Mountains close to Nordhausen and the southern one was to be situated between Vilvna-Neustadt and Friedrichshafen.

V-2 rocket on its mobile launcher. (Author)

V-2 rocket lifting off from Peenemunde. (Author)

The Fieseler Fi-103 had been designed to be either ground launched or air launched from bomber aircraft. On 22 August 1942, the activities at Peenemunde attracted the attention of British Intelligence after a Fieseler Fi-103 V-1 Reichenberg had crashed almost intact into a turnip field on the island of Bornholm, Denmark. It was thought that the crashed Fieseler Fi 103, with a dummy one-ton warhead made of concrete, had been dropped from a Heinkel 111 during a test flight from Peenemunde, but its pulse-jet motor had failed and it had come down on the island. The missile had gone off course and landed close to the shore of Bornholm. Within minutes the local

V-2 rocket being prepared for launching on the island of Peenemunde. (Author)

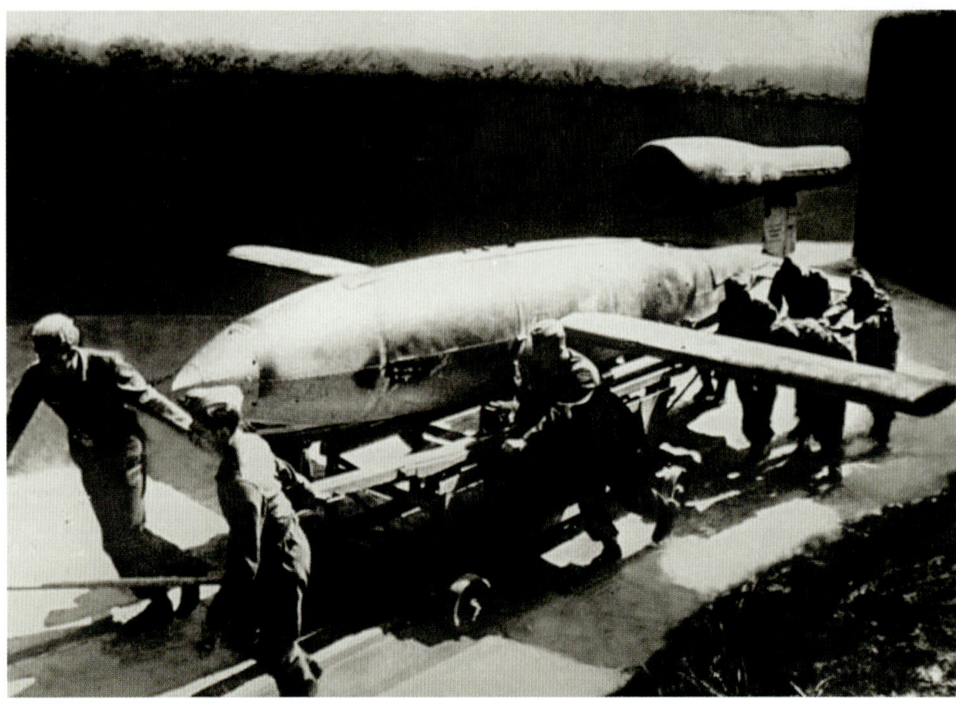

V-1/flying bomb being manhandled to launch site. (Author)

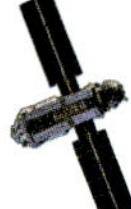

Wreckage of a V-1 flying bomb being examined after crashing on the island of Bornholm. (Author)

policeman, Johannes Hansen, and a naval intelligence officer, Lieutenant Commander Hassager Christiansen, had arrived. Hansen took photographs of the missile whilst Christiansen drew a number of sketches.

A very short time later a German officer accompanied by soldiers arrived and cordoned off the area. The German officer asked Christiansen if he had taken any photographs of the missile, to which Christiansen replied, 'No'. This was of course true because Hansen had taken the photographs and he was not asked. The missile was recovered later by a team from Peenemunde under strict security and taken away, but within days copies of the photographs and sketches of the missile were on their way to London. Unfortunately the package containing most of the photographs and sketches was discovered aboard the Danish ferryboat that went to Sweden. The Germans realised that there was only one person who could have taken the photographs and that was Lieutenant Commander Hassager Christian. He was arrested and tortured by the Gestapo, but he revealed nothing and was imprisoned in a hospital awaiting execution. He was rescued by the Resistance who then spirited him away to Sweden where he worked with the intelligence section there until the end of the occupation. Fortunately some of the photographs and drawings did manage to get to London and were invaluable in locating the launch sites. Some months later the remains of a V-2 rocket engine were recovered from the sea off Bornholm by members of the Danish resistance.

Part of a V2 rocket being recovered from a lake during WW2. (Author)

On 17 August 1943, Operation Crossbow, a 600-bomber raid by the RAF, pounded the Peenemunde research establishment, destroying hangars, laboratories and launching facilities and causing over 800 casualties. However by the end of September mass production had started again, though dramatically reduced. The Cherbourg base had been captured by the Americans, the eastern plant in Latvia was in Russian hands and the site in the south had been so badly damaged by allied air raids that it was considered to be almost ineffective.

As the Russians advanced and the Germans retreated, senior officers of the Waffen SS (Schutzstaffel) ordered that all the technicians and engineers from Peenemunde and the other rocket facilities be rounded up and taken to Munich and executed, to prevent them falling into the hands of the Allies. Fortunately the speed of the Allied advance was such that the convoy carrying them was intercepted and the technicians and engineers taken prisoner. Also all the blueprints and documents from Peenemunde were loaded onto trucks by the Waffen SS and were being taken to be stored in an abandoned mine in the Harz Mountains, but they had been spotted by the advancing American forces, who subsequently recovered them.

With the Allies now in Germany and the German army in total disarray, the war was essentially over and the race started to find the rocket specialists. Magnus von Braun, brother of Werner von Braun, made contact with the Americans and, after proving who he was, surrendered himself to the Allies, together with his brother and 150 scientists, engineers and technicians. The race was now on to obtain the information for the development of long-range rockets and guided missiles. This information was desperately sought after by the Russians and the Americans and

tremendous efforts were made to capture the scientists and engineers involved. But it was the Americans who managed to capture the main people involved – Werner von Braun, Hermann Oberth, Walter Dornberger and many others. I say captured, but it was Oberth and Dornberger who sought out the Americans, rather than give themselves up to the Russian Army.

The Second Belorussian Front, commanded by Marshal Konstantin Rokossovsky, entered Peenemunde on 5 May 1945 and after taking control, started to gather up all that was left of the rockets and related equipment, including what engineers and scientists were left. Amongst some of the items they found were the blueprints of the A9 and A10 inter-continental rockets, the V-2 blueprints, the V-1 and a number of surface-to-surface and air-to-air experimental missiles, R-4M, R-10 and R-12 rockets amongst a number of others. An engineer from the NII-1Research Institute, Alexi Isaaev, who was amongst the first to arrive in Peenemunde, discovered a document amongst the rubble that described a rocket-propelled, supersonic bomber in great detail. But the vast majority of the information and personnel were already in the hands of the Americans. Despite this setback, Russia managed to capture Helmut Groettrup, an electronic specialist, and Erich Apel, a rocket engineer, who then in turn persuaded a number of specialist German engineers to join with them. The Russians decided to reactivate Peenemunde and within months, Helmut Greottrup and the other engineers had a V-2 production line running, along with limited laboratories and test facilities. However it soon became obvious that so much damage had been done to the facilities at Peenemunde during the 600-aircraft bombing raid, that although it appeared to be up and running, it was in fact struggling and the project was finally abandoned. What had been achieved by the Russians at Peenemunde was systematically destroyed before they left.

The Soviet army entered Thuringia on 14 July 1945 and headed towards the German rocket centre at Nordhausen, home of A-4 rocket production. After capturing this, Russian specialists, including Boris Evseevich Chertok, who had examined a captured A-4 rocket in Poland, arrived in Nordhausen to examine the facilities there. They discovered a similar situation to that of Peenemunde: all the key personnel, together with rockets, blueprints, diagrams and other documentation, had gone. The Russians entered the underground factory and discovered a large number of A-4 rocket parts and other electronic pieces. There were also a number of German rocket engineers still at Nordhausen, who had declined to leave and who provided the Russians with a great deal of information about the facility. In return for their co-operation they were offered a far better lifestyle, good wages and comfortable accommodation.

With very little to be gained by staying at Nordhausen, Boris Chertok and Aleksey Isaev, together with twelve German engineers, were sent to Bleicherode, the last headquarters of Werner von Braun. Within days of arriving, Chertok had created a new institute called the RABE derived from the Raketenbau und Entwicklung (Rocket Building and Design), which was given the task of restoring the flight control system of the A-4 rocket, the most complicated part of the whole program. With news coming through that Lehester, a town near Bremen, had been vacated after being captured by the Americans, two of the engineers from Peenemunde, Aleksei Isaev and Arvid Pallo, were immediately sent there. This was one of the very few rocket-related facilities that had been found completely intact and was the site of the test firing stand for the A-4 rocket engine. The Russians also sent their top rocket engine specialist, Valentin Glushko, together with experts from the OKB-SD design bureau, who specialised in designing systems for rocket-assisted military aircraft. The design team immediately set to work with Glushko in reactivating the site for the resumption of test firings.

At the end of the Second World War the remains of some Fieseler V-1 flying bombs had been delivered to Moscow from Poland after the Russians had overrun the test range at Blizna, and Vladimir Chelomei was asked if he could build a similar one. He said yes, and found himself in charge of a 100-strong engineering and design team. Within months he and his team had produced an almost exact copy and by the end of 1945 had produced a better version of his own design. The 10Kh or Izdeliye, as it was known, underwent numerous tests but the guidance system became a continual cause for concern and the development was dropped. Over the next decades Vladimir Chelomei designed numerous types of missiles and rockets that included the Proton launch rocket and later the Almaz space station.

Some months later, in August 1946, the RABE institute was placed under the control of the GAU (Chief Artillery Directorate), although Chertok was at first reluctant, as he realised that most of Russia's bureaucrats were not interested in rocket technology. Fortunately the military were and they offered support and official recognition after a visit to Bleicherode by General Ivanovich Kuznetsov and General Mikhailovich Gaidukov, who was chairman of the production department that oversaw the building of the Katuysha rocket. Impressed with what he saw and been told, Gaidukov decided to use his personal contacts within the Central Committee of the Communist Party to lobby for more support for the development of Soviet rocketry.

Joseph Stalin, not known for his patience, was dissatisfied with the progress being made and ordered that the whole project be shut down and everything moved to the Soviet Union. Within days of the order, known as Operation Osoaviakhim, being given, General Ivan Serov, head of the KGB (Komitet Gosudarstvennoi Bezopasnosti), had more than 6,000 German engineers and technicians and about 20,000 members of their families unceremoniously put on to trains and taken into the Soviet Union. They were then shipped to various research facilities throughout Russia. A large number were sent to the Kubyshev rocket center at Samara where, under the charge of Yuri Pobedonostev, a former member of the TsAGI and RNGII, they carried out a series of tests and experiments on the V2 rocket with the intention of improving it and putting it into full production. By the end of 1945 production was in full swing and had been improved dramatically. In Lehester they ceased test firing operations in January 1947 and destroyed the test rigs before leaving. Then, together with Russian scientists and some of the German scientists and engineers already in Russia, they concentrated on developing their missile and rocket program. A new test site was developed in 1947 at Kapustin Yar, which became the cradle of Soviet rocketry when Soviet engineers and their German colleagues launched A-4 ballistic missiles from this dusty site on the banks of the Volga River.

CHAPTER TWO

Into the Space Race

Russia had always been reticent about disclosing its progress and way of life, a situation that had got worse since Joseph Stalin had become leader and created a totalitarian country. Now with the war over and countries licking their wounds, rebuilding their lives and repairing the almost insurmountable damage they had inflicted on each other, Russia retreated farther and farther back inside its own world. There was a total blackout on news both in and out on anything concerning the military and the development of the rocket program. Scientists who were working on these projects remained anonymous and anything that was published was scrutinized and vetted, which in fact meant that they revealed almost nothing.

The Minister in charge of armaments, Dmitry Fedorovich Ustinov, was given the responsibility of developing the rocket and missile program. He had been impressed with Korolev's expertise in the field of rockets as well as his organizational abilities and appointed him the Head of Construction for the Development of Long-Range Missiles. With this appointment came the rank of Colonel in the Red Army, but technically Korolev was still a prisoner and subject to the rules governing prisoners.

Korolev ordered the immediate roundup of 200 German workers from the Mittelwerke V-2 factory and had them shipped to Lake Seleger, which was situated between Leningrad and Moscow. In a strange twist of fate, the prisoner now became the jailer and responsible for the security of the facility and its inmates. The majority of the Germans had no direct contact with the construction of the rockets or the Russian engineers building them, all their input came from interrogation and written sources.

In the early years it was thought that the development of the Russian space program was for propaganda purposes and to show the world the strides communism was making. But those who knew Sergei Korolev were well aware of his ability, dedication and the progress he was making in this field. For many years, Korolev's name had been kept secret, and only those who needed to know knew that he was the head of the space program.

Because the war had left the Soviet Union bereft of essential materials usually associated with the manufacture of rockets, Korolev used stainless steel in place of aluminum and titanium, making the construction of his designs heavier than the American ones. This meant that the engines had to be more powerful, and instead of developing large complicated engines, he developed much simpler engines mounted in several pods around the rocket.

There were also other problems. The engineers selected for the rocket program came from the military Strategic Rocket Force and, unlike the majority of their American counterparts, who in their youth had tinkered with engines and radios, the vast majority of Russians had no experience of this. There was a singular lack of automobiles and radios in the Soviet Union in the early years and this prevented young men and women from gaining any form of experience. Therefore the rockets developed had to be of basic design and construction, rugged and reliable to enable them to endure the unsophisticated hands of the engineers and the harsh winters and extremely hot summers of this vast country.

In 1949 the Russians had developed their own version of the V-2 rocket, known as the Pobeda and named after its designer General Andrey Pobedonostev. It had a range of 560 miles, far greater than that of the German V-2. It formed the catalyst for the formation of the first Soviet Rocket Division, armed with V-2 and Pobeda rockets.

From the rockets that launched the missiles, came the first tentative steps to putting a rocket into space and ultimately a man into orbit around the Earth. In 1949 the Russians launched a single stage rocket from Kapustin Yar, Astrakhan Oblast, near Volgograd that could reach an altitude of sixty-eight miles and was used for geophysical surveys. Carrying a payload of instruments weighing 280lb, it obtained valuable information on the atmosphere at these maximum heights. The rocket later descended to Earth on a parachute, enabling the scientists to recover and evaluate the instruments. So confident were the Russians in the development of these rockets that five more were launched (only two were successful) and, at a meteorological conference in Washington DC, they submitted a technical paper in which they described a two-stage rocket that they had developed. It consisted of a twenty-three feet long sustainer rocket powered by a nitric acid and kerosene engine, developing a thrust of 3,014lb for sixty seconds assisted by a solid-propellant booster rocket. At an altitude of forty-three miles the sustainer rocket separated into two sections, the upper section carrying the instruments continuing to a height of sixty-eight miles to carry out its surveys. Both sections were later recovered by parachutes. The rocket was later identified as an MR-1 that was capable of carrying a payload to an altitude of 120 miles.

The first Soviet-built rocket, the R-1, was almost a direct copy of the German V-2, but built with Soviet components. This was supposed to be the first of the Russian ICBMs, but the R-1 only had a range of 3,500 miles. Because this rocket could barely reach the United States, the development of the next phases, the R-2, -3, -4, -5 and -6, was hurriedly brought forward, as the tension between the two super powers increased.

The first tests of the Soviet-built V-2 (R-1) took place at Kapustin Yar 13 September 1948. The first two test rockets were successfully launched and travelled close to 200 kilometres, however the first rocket landed thirty kilometres away from its intended target, whilst the second rocket landed 180 kilometres away. Two of the thirteen German engineers who were present at the tests were Johannes Hoch and Kurt Magnus, specialists in guidance systems; they recovered the wreckage and after some intensive readjustments and tests, they solved the problem. The guidance system had been a problem with the V-2 since its early days at Peenemunde. Then in July

1947 a team led by Helmut Gröttrup put forward a proposal for an improved version of the V-2 (G-1) or R-10 as it was known in Russia. However, although supported by some senior managers, it was vigorously opposed by Sergei Korolev, the chief designer of long-range ballistic missiles, because he had already submitted a design for an improved version of the V-2 called the R-2. He also objected to the use of German specialist engineers on such a sensitive project. Valentin Glushko, who had just been appointed chief designer of liquid-propellant rocket engines and had been using German engineering specialists to develop a copy of the V-2 called the RD-100 and later the RD-103, was of the same opinion. He decided that he too no longer required German engineers to work on what was developing into a sensitive and secret project. The Russian secrecy mentality started to influence their thinking and the decision was made that they no longer needed German expertise in building the long-range missiles, so all the remaining Germans were banned from all the sites and facilities that were actually building the rockets. Their expertise was still to be used, but to be confined to design and development and well away from the manufacturing facilities.

In the 1950s, the first of a series of test flights of the R-1 rocket were made, each carrying live animals. The first on 15 August 1951 carried two dogs, Dezik and Tsygan, who were the first to make a sub-orbital flight when their rocket reached an altitude of sixty-eight miles (110km). Both dogs were recovered unharmed. Later that year Dezik, together with another dog called Lisa 1, made another sub-orbital flight, only this time both dogs died after the parachute on their capsule failed to deploy. Tests using animals continued and on 26 July 1954, two dogs, Lisa 2 and Ryzhik, were launched on an R-1 to an altitude of sixty-two miles (100km). It is not known whether or not they survived but it is unlikely. At the beginning of September 1961 two more dogs, Smelaya and Malyshka, were launched on a sub-orbital flight and although their spacecraft crashed on re-entry after their parachute had failed to open, they both survived. This was followed in the late September by the launching of two more dogs called Bobik and Zib, who made a successful flight to an altitude of sixty-eight miles (100km) and were recovered successfully. The first test using an ejection procedure was carried out using two dogs, Albina and Tsyganka, who after a sub-orbital flight were successfully ejected in their capsule at a height of fifty-three miles (85km). Both dogs were recovered and survived. In December 1960 another mission that was part of the Vostok program was to put two dogs 'Damka and Krasavka' and a number of mice, into orbit. After a series of equipment problems, the mission was aborted when at a height of 133 miles (214km). The spacecraft re-entered the atmosphere and at a predetermined height was set to eject the two dogs and then self-destruct. Unfortunately both the ejection seat and the self-destruct mechanism failed to activate. The capsule landed in deep snow with a self-destruct back-up mechanism primed to activate in sixty-five hours. At a temperature of -45 degrees F, the rescue teams sent to the area to recover the dogs found no signs of life and it was too dark to attempt any form of rescue. In the morning they heard barking and on opening the capsule, and making safe the destruct mechanism, discovered both dogs alive and well, although all the mice had died from the extreme cold. Both dogs were immediately wrapped up in warm blankets and whisked away to

Space dogs Veterok and Ugoljo. (Roscosmos)

Moscow. Sergei Korolov wanted to make the story public, but such was the continuing secrecy surrounding the space program, that the authorities refused. Krasavka was later adopted by a leading Russian physicist and his family and lived for another fourteen years as the family pet. All the dogs that were used in the space missions and survived ended up as the family pets of people concerned in the development of the rockets or the space program itself. There were a number of other sub-orbital flights using dogs and different animals, some successful, some not.

Whilst the experiments with animals had been going on and with the success of the ballistic missile program, the Russians turned their attention to expanding their biological and meteorological experiments and surveys. However they were still working on designing and building bigger and better ICBMs (Inter-Continental Ballistic Missile), then in 1957 a design was submitted by Korolev's group for consideration and was accepted. Such was the progress, that in just under two years a number of ICBM rockets were ready for testing. The improvements of the R series of rockets continued and the R-7 was produced and by the beginning of 1953 was ready for testing. It was a complete success, and the threat to world peace stepped up a gear as the world's first ICBM was announced to the world. This of course was a worrying time for the west and galvanized the United States into increasing their rocket capability.

Sputnik 1 being launched from Baikonur. (Roscosmos)

Then came news, reported in *Pravda*, that there had been a successful test of an 'Intercontinental Ballistic Rocket' together with tests of nuclear and thermonuclear weapons, all claiming to be in the interests of research. Such was the secrecy in Russia, even to their own people, that it wasn't until the very end of September that it was announced that it was about to launch two different types of satellite and gave out the frequencies so that people could listen to the 'bleep' signal as they passed overheard.

Because of the increasing number of launches now being carried out, it was decided to build a purpose-built launching site, and the place chosen was outside the small town of Baikonur, Kazakhstan, which at the time was part of the Soviet Union. Work began in 1955 and two years later was completed, and it remains to this day Russia's main Cosmodrome.

Then on 4 October 1957, Baikonur Cosmodrome trembled as an R-7/SS-6 ICBM rocket, carrying a 23-inch (58.0cm) diameter metal ball, weighing 184lb, with four 2.9m whip-like antennas protruding from it, perched inside its needle-like nose and fitted with a radio transmitter, was blasted into space. At a height of 142 miles and at a velocity of 17,890 mph, the engine was shut down and the protective nosecone jettisoned. The satellite, *Sputnik 1*(Fellow Traveller), as the tiny object the size of a basketball was called, then separated from the rocket and began to free flight around the Earth reaching an apogee of 588 miles. The moment the satellite separated, the 9ft (2.9m) antennas unfolded and two radio transmitters on frequencies of 20.005 and 40.002 MHz were activated and started

Sputnik 2 being prepared for launching. (Roscosmos)

Russian space dog Laika in her space bed. (Roscosmos)

transmitting. Minutes later a faint bleeping sound was being tracked around the world, as *Sputnik 1* orbited the Earth. The carrier rocket orbited the Earth for sixty-two days, whilst the satellite *Sputnik* orbited the Earth for the next ninety-four days.

The sphere was filled under pressure with nitrogen, which was intended to test for meteoroid detection. In the event of a meteoroid hitting the sphere and penetrating it, the loss of pressure would have been recorded in the temperature data being transmitted back to Earth. No such incidents were recorded. One of the experiments was carried out by observing the decay rate of the carrier rocket before it re-entered the Earth's atmosphere. An important experiment was carried out investigating the ionosphere using the 20.005 and 40.002 MHz radio waves being emitted from the satellite. Radio waves of these frequencies are propagated through the ionosphere, not in a straight line, but in a curve, and measuring the time between the satellite's optical and radio rising and setting made it possible to measure the curve of the linear distortion of the radio beam, thereby determining the electron content of the atmosphere in the path of the beam. The results of this experiment were extremely important for ensuring the reliability of communication between spacecraft and mission control on future missions. The chemical batteries on board the tiny spacecraft powered the radio transmitters for three weeks before finally shutting down. *Sputnik 1* carried out over 1,400 orbits of the Earth before going into a decaying

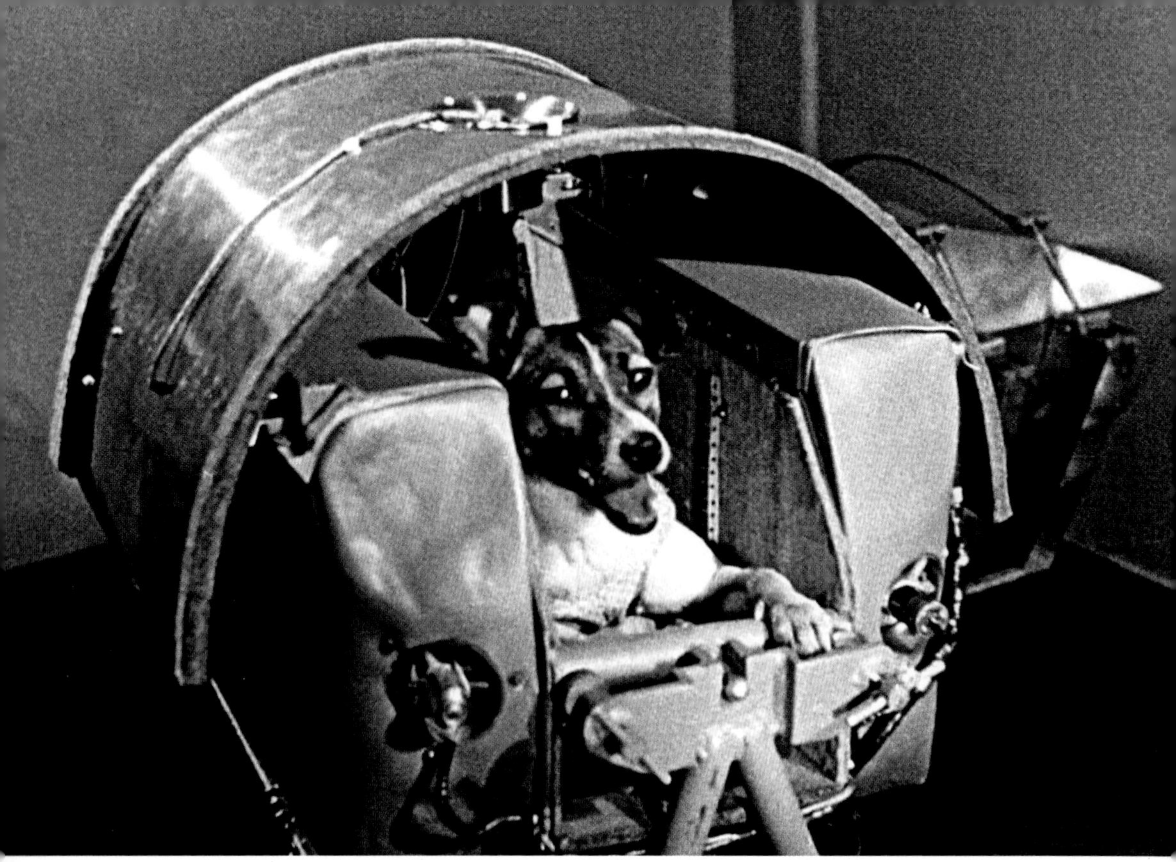

The Russian dog Laika in her space compartment. (Roscosmos)

orbit and burning up during re-entry on 3 January 1958. But this mission had more far reaching effects on the world: man had placed an object in space and controlled it, and the race to attempt to conquer this element had started in earnest.

This significant step came at the height of the Cold War that existed between East and West and caught the world completely by surprise. In the United States there was a mixture of shock and disbelief, as they had been led to believe that the Russians were way behind in their technology. Their immediate thoughts were of military satellites now spying on them, followed by the possibility of missiles raining down upon them from space.

On 3 November 1957, one month after *Sputnik 1* had been launched and stunned the world, another R-7/SS-6 ICBM rocket blasted *Sputnik 2* off the launch pad at the Baikonur Cosmodrome at Tyuratam, and the first living being to orbit the Earth was launched into space. The living being was a part-Samoyed terrier dog named Laika. She had been a stray dog found wandering the streets of Moscow and had been trained for a spaceflight role.

Laika was placed in a pressurized padded compartment that enabled her to either stand or lie down, but all the while a harness supported her. Her food and water was dispensed in gelatin form. Separate compartments held the air regeneration and temperature control systems, the radio transmitters, telemetry and scientific systems. From the onset it was realised that there was no provision in place for returning the dog safely to Earth, so the plan was to poison her with a serving of food. The authorities said at first that she had spent seven days in orbit before being put to sleep by means of a portion of poisoned food. But it is now known that she died a short time after the launch from heat and distress, after part of the nosecone was ripped off during the launch.

Despite her early demise, Laika had provided scientists with the first, but limited, data on a living organism in a space environment. The tiny spacecraft orbited the Earth for the next 162 days, before going into a decaying orbit and burning up on re-entry on 14 April 1958. One of the most important experiments carried out was the discovery that as the spacecraft entered the polar regions of the outer zone of the Van Allen Belt, the orbital altitude and geomagnetic latitude increased and there was also an increase in the flux intensity of the high-energy charged particles. In 2008 a small monument to Laika was placed outside a military research station outside Moscow, as a tribute to the little dog who gave her life in the furtherance of space exploration.

The Americans, initially stunned by the sudden progress made by the Russians, stepped up their development program and launched their first artificial satellite, *Explorer I*, into orbit on 1 February 1958. It weighed 8.3 kilograms and had a diameter of only 15cm. This was closely followed on 15 May 1958, when the Russians launched an unmanned automated space laboratory, *Sputnik 3*, on top of a modified R-7/SS-6 ICBM rocket just over three months behind the Americans. Plans as far back as July 1956 had been drawn up to place the world's first artificial satellite laboratory in space, but Soviet bureaucracy caused the program to slip further and further behind. The automated laboratory payload weighed 2,129lb and contained twelve instruments, which provided data on the composition of the upper atmosphere, measured the magnetic and electrostatic fields and the photons in cosmic rays, amongst other things. The instruments were also designed to measure the corpuscular radiation of the Sun, which discovered electrons with an energy of 10-KeV for the first time. Studying these electrons showed that they were probably atmospheric electrons that were accelerated in the outer atmosphere by the transitory geomagnetic fields. The experiment was also set to map the Van Allen radiation belt, but the tape recorder failed to operate and so no information regarding this was sent back. Almost two years later and after over 10,000 orbits of the Earth, *Sputnik 3* re-entered the atmosphere and was burnt up.

The development of these simple rockets had a knock-on effect. Their lack of technological experience caused the Russians to launch large numbers of rockets, whereas the Americans, with their more sophisticated rockets, launched fewer, but were able to acquire a similar amount of information. The Russians relied for many years on batteries to power their spacecraft, whilst the Americans had progressed to using solar power. This meant that the Russian spacecraft spent much shorter periods in space, which in turn meant that they had to launch considerably more rockets. Because of this they gained vast experience in launching and recovery techniques, whilst in this field the Americans lagged behind.

Both the Americans and the Russians continued to make progress towards the dream of putting a man into space, but it was the Russians who were to make the first breakthrough. President Eisenhower at the time refused to be drawn into a space program based solely on military grounds, but he did approve the setting up of a civilian agency that looked at the future of space travel – NASA (National Aeronautics and Space Administration), which replaced the NACA (National Advisory Committee for Aeronautics) that had been established in 1918.

In Russia, Sergei Korolev had been working hard on improving the R-7 launch vehicle that had been used to launch the Sputniks. The problem that he had was that, unlike the Americans, there was no central bureau to oversee the design, development and building of rockets. The Russian Premier at the time, Nikita Khrushchev, refused to centralize the space program and left it to various ministries to control. This meant of course that Korolev had to delegate the majority of his work to trusted members of his staff, whilst he kept track of the work being done by the other ministries, which ultimately slowed the process down considerably. Amongst the trusted members of staff was Korolev's deputy, Vladimir Nikolaevich Chelomei.

Vladimir Chelomei was one of Russia's leading rocket designers in the formative years of space exploration, together with Sergei Korolev and Mikhail Yangel. Chelomei first came to the fore when he headed up the OKB-52 design bureau, which designed and constructed a range of long-range cruise missiles and spy satellites.

Born in 1914 in Sedletse, Poland into a family of teachers, Chelomei obtained his Master's degree in engineering at the Kiev Aviation College. He was awarded the Joseph Stalin Doctoral Scholarship soon afterwards where he studied pulse-jet engines. Vladimir Chelomei had been appointed head of the Special Design Bureau OKB-52 in 1955, which was heavily involved in the design and construction of missiles and was slowly moving into the design of space rockets. He shrewdly hired Sergei Khrushchev, son of the Russian Premier and a guidance systems engineer. With the backing of the head of the government, he soon had the largest budget of any bureau in the Soviet Union, but with none of the expertise. This was soon put right, when Valentin Glushko, head of the Gas Dynamics Laboratory (GDL) and principal designer of Russian rocket engines, joined forces with Chelomei after he had had a serious falling out with Korolev.

The animosity between Korolev and Glushko was not new. In the 1930s it was Glushko's testimony that had been partly responsible in getting Korolev sent to a labour camp during the Stalinist purges. In the 1950s, Glushko had been Korolev's deputy, and the two had had major disagreements regarding the use of high-energy cryogenic fuels. Korolev wanted to use the new fuels, whilst Glushko wanted to retain the use of the highly toxic hypergolic chemicals that ignited on contact.

Chelomei, on hearing of the disagreements, quickly approached Glushko with regard to using his engine in one of his designs for the UR-500 Proton ICBM rocket. With the backing of Nikita Khrushchev, Chelomei was given the go-ahead to design and build a manned spacecraft for a circumlunar flight – the LK-1 project.

Vladimir Korolev had submitted two plans for the design of a manned spacecraft. The first was for a three-man spacecraft, which would eventually become the Soyuz model, that consisted of four modules. A service module, living module, a command

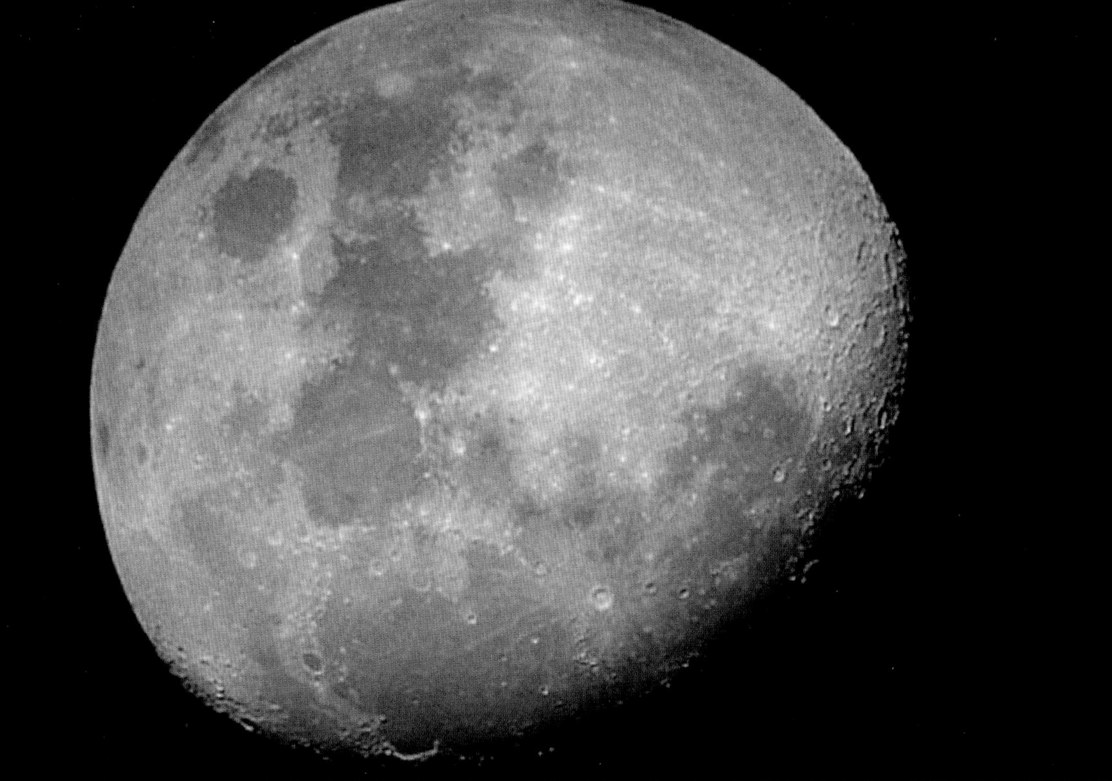

Photograph of the Moon taken by *Luna 1*. (Roscosmos)

and re-entry module and an attitude control module. Once in orbit the modules would transpose and dock with the L-1 rocket and then proceed to the Moon.

The second proposal, known as the Soyuz complex, consisted of three spacecraft given designations A, B, and C. The Soyuz A model was in fact the first proposal with the four modules, the B model was an unmanned version of A, whilst the C model was an unmanned fuel tanker for refueling model B whilst in orbit. Both proposals were submitted to the government committee and both were rejected, mainly because of Sergei Khruschev's influence and his association with Chelomei.

Despite claims to the contrary, both Russia and the United States entered into a race to be the first to send a manned spacecraft to the Moon. The Russian program, the Lunar Circumnavigation Mission, started when their unmanned spacecraft *Luna 1* was launched on 2 January 1959. It had been programmed to orbit the Moon and send back information, but the battery powered sphere raced past the Moon at an altitude of 3,660 miles (5,891km) and then went into a fifteen-month decaying orbit around the Sun. Despite missing its target, it did manage to send back information showing that there was no perceptible magnetic field close to the Moon, but it did detect however that there were streams of ionized plasma in space also known as the solar winds.

On 12 September 1959 the Russians launched *Luna 2* on a Luna 8K72 rocket which crashed into the Moon's surface near the crater Archimedes, becoming the first man-made object to land on another celestial body. Although the spacecraft crashed, it did manage to transmit data back to Earth showing that the Moon did not have a radiation belt around it. Just prior to impact two sphere-shaped balls were released from the spacecraft, each with seventy-two pentagon-shaped pieces in a pattern similar to that of footballs. Each segment was made of titanium alloy and was emblazoned with the emblem of the Soviet Union and the letters CCCP (USSR) in Cyrillic engraved beneath. As the spheres made impact with the lunar surface, an explosive charge inside was activated sending the segments in different directions.

Then on 4 October 1959, to celebrate the second anniversary of *Sputnik I*, the Russians launched *Luna 3* on a Luna 8K72 rocket from Baikonur; this time the tiny spacecraft was fitted with a camera and programmed to go into orbit around the Moon. As it disappeared around the back of the Moon, it re-appeared, after what was felt like an almost an interminable silence (40 minutes), beaming back pictures of the previously unknown far lunar surface. *Luna 3* took a total of 29 images covering seventy percent of the far side of the Moon. As the spacecraft returned to Earth, efforts were made to transmit the images back, but because of a low signal strength were unsuccessful. However, as the spacecraft neared Earth another attempt was made and this time managed to recover seventeen images, albeit of a poor quality. Contact was lost with the spacecraft soon after and it is believed to have burnt up on re-entry. Despite the poor quality of the images, this was a remarkable achievement, but it was to be Russia's last major success for the next two years. A further four attempts were made to launch a Luna spacecraft to the moon, but all failed and it wasn't until 1963 that the next successful mission was launched.

Cosmosphere taken to the Moon aboard *Luna 2*. (Roscosmos)

Luna 3 flyby probe. (RKS Energia)

Korolev, in the meantime, had been working on his own project, known as the N (Nositel) series. Despite not having the political backing or the funding for full-scale development that Chelomei had, Korolev and his design team set about developing the N-1-L3, but it was turned down in favour of Chelomei's LK-1 rocket. Korolev's original plan was for three N-1 boosters to place a 200-ton payload into Earth-parking orbit, which in turn would launch the L-3, with a payload of 75 tons, to the Moon for a landing and subsequent return. Korolev was offered funding for the design and development of four N-1s, but he maintained that for that kind of money he could only carry out a lunar orbit. The consensus of opinion by the government was that the N-1 was incapable of carrying out all that Korolev claimed, and that the 200-ton payload was beyond the capability of the engines to place in orbit.

Later, both Korolev and Mishin admitted privately that they had miscalculated the minimum payload and that the Russian engines were nowhere near as good as the Americans. If they had admitted this in public the whole project would have been killed, and they were desperate to build the N-1 at any cost, even if they knew it would result in its subsequent failure.

On 2 April 1964, an ambitious mission was started when *Zond 1* was launched on a Molniya 3 rocket, to carry out a flyby of the planet Venus. Unfortunately a cracked sensor window caused the spacecraft to depressurize, causing a major loss in radio communication. With almost all control lost, the spacecraft missed the planet by 100,000 miles and was lost. Eight months later on 30 November 1964, *Zond 2* was launched on a Molhiya T-103 rocket, this time to carry out a flyby mission of the planet Mars. Once again communication problems caused contact to be lost and the spacecraft flew past the planet Mars and was lost.

In August 1964, the Central Committee issued an order that a full lunar program was to go ahead. Chelomei's LK-1 was to be developed for a manned lunar flyby, whilst Korole-Molniya 3v's N-1-L3 was to be developed with the intention of carrying out a manned landing on the Moon. The proposed lunar landing missions were to be carried out in 1967/68

Korolev was forced to abandon developing his 75-ton military space station named *Zvezda*, armed with nuclear missiles and ready to fire from orbit. The project never materialized, but the *Zvezda* space station concept was developed many decades later to become part of the ISS.

There were problems right from the start, the main one being the time-scale laid down by the Central Committee. There were other problems not associated with the program and that was the in-fighting that was going on between the designers Korolev, Chelomei and Yangel. One of the most influential scientists around at the time was a mathematician and aerodynamicist by the name of Mstislav Keldysh. He tried to play the diplomat by supporting the three main protagonists in areas that he knew was their expertise. He supported Yangel for his work with ICBMs, Korolev for the N-1-L-3 and Chelomei for his UR-500 LK-1. Korolev then decided to concentrate his efforts into the design and construction of a new and advanced spacecraft, the result being the Soyuz (Union) multipurpose vehicle. The design had been chosen

to replace the Vostok manned spacecraft which, although reliable, did not have the capability of being able to change orbit in space or rendezvous and dock with other spacecraft, and could carry only one cosmonaut.

Because of his falling out with Glushko, Korolev had to look for another engine designer/manufacturer and joined up with Nikolai Kuznetsov. The problem was, that Kuznetsov had only designed and built aircraft engines and had no experience with rocket engines with the power needed to put a rocket into space. Despite this his team got to work and managed to produce an engine, albeit with very limited power, to be used in the early trials. When the N-1 was reintroduced in 1969, as the launch vehicle for the proposed Soyuz spacecraft, it took 30 of the engines in the rocket's first stage to put it into orbit.

The Soyuz spacecraft consisted of three modules: the command, service and orbital. The Command or descent module housed all the radio equipment and instruments for controlling the spacecraft's descent, the life support systems and the food and water supply. Directly in front of the pilot was the spacecraft's control panel on which were all the instruments for controlling the on-board systems and equipment, navigational equipment and a television screen. Either side of the cosmonaut's contoured couch were two levers that controlled the spacecraft. The left-hand one altered the spacecraft's speed whilst manoeuvring, whilst the right-hand lever controlled the yaw, pitch and roll.

The Service module contained the bulk of the on-board equipment and propulsion systems, all of which were in a sealed instrument container. The long-range radio and telemetry equipment were also housed in this compartment. The liquid propulsion unit used for manoeuvring the spacecraft and for re-entry, together with the propulsion engines, was contained in a non-pressurized section of the module. Also housed in the module was the centralized electrical-power system which was supplied by solar battery panels. These were recharged by rotating the spacecraft about its axis to ensure that the panels were constantly pointed toward the sun,

The Orbital module was the scientific section of the spacecraft but was also used for recreation purposes, physical exercise and sleeping. Two of the sections, the Orbital and Service modules, were separated from the descent module prior to re-entry. On re-entry the descent module was traveling at 755 feet per second and had to be slowed to about 5 feet per second for landing. This was achieved by the deployment of four parachutes just fifteen minutes before touchdown and then released. This was followed immediately by two more parachutes that slowed the descent speed further. These were then released and followed immediately by the main parachute, which had a surface area of 10,764 square feet, slowing the spacecraft to 24 feet per second. Seconds before touchdown two sets of three descent engines on the bottom of the spacecraft were automatically fired, bringing the spacecraft to a soft landing.

In the United States, the Americans had made another step forward in the 'space race' when, on 9 April 1959 they selected seven experienced military test

pilots to train as astronauts. They were to be known as the Mercury 7. Not to be outdone, the Soviet Union too had started to select cosmonauts. A large number of Soviet military personnel applied for selection, but this was quickly whittled down to six. The first six Soviet cosmonauts were selected in May 1963: Pavel Popovich; Yuri Gagarin; Valentina Tereshkova; Valeri Bykovsky; Andrian Mikolyev and Gherman Titov. All the selected cosmonauts were awarded the Hero of the Soviet Union. During the initial selection two cosmonaut-candidates were dropped almost immediately due to accidents: Valentin Varlamov damaged his spine in a diving accident, whilst Anatoli Kartshov suffered a hemorrhage of the spine during training on the centrifugal force machine in which he experienced over 8g.

The Russian criteria for selection differed from the Americans, inasmuch as the height of the cosmonauts was restricted to a maximum of 5ft 7ins and a maximum weight of 154lbs (11 stone). Their flying experience was extremely limited when compared with that of the Americans, i.e. Yuri Gagarin had only 240 flying hours, Alexi Leonov 250 hours and Gherman Titov 240 hours, whereas the Americans were all military test pilots and had to have had a minimum of 1,500 hours each. Valentina Tereshkova,

Some of the first Russian cosmonauts: back row, left to right: Leonov, Titov, Bykovsky, Yegorov, Popovich; front row: Komarov, Gagarin,Tereshkova, Nikolayev, Feoktistov. (Gromov Institute)

Cosmonauts Pavel Popovich, Yuri Gagarin, Valentina Tereshkova with Sergei Korolev outside the Cosmonaut Training Centre. (Roscosmos)

although not a pilot, was an experienced skydiver which, from the Soviet perspective, was a must. The initial cosmonaut tests started with each candidate having to spend several weeks in an isolation chamber, then being placed in a thermal chamber for a period of time, followed immediately by a similar period in a decompression chamber. On completion of these tests the candidate was placed aboard an aircraft and had to parachute out. The reasoning behind these tests was to discover if the candidate could deal physically and psychologically with the rapid transition from being in a confined space to being released into boundless space within a short period of time. All these tests were dangerous as they had never been tried before, but the results were to help set the yardstick by which all future cosmonauts would be selected. An example of the dangers was when a young 24-year-old trainee cosmonaut, Valentin Bondarenko, who had taken the place of Anatoli Kartchov, had just completed his initial tests satisfactorily, and was then placed in the isolation chamber with its pure oxygen atmosphere. He had just completed this phase when the atmosphere inside the chamber ignited, turning it into an inferno. It was discovered later that Bondarenko had removed his biosensors and wiped his body with cotton wool and then thrown the wool carelessly onto an electric heater. The cotton wool caught fire and ignited the oxygen-rich atmosphere. Bondarenko died eight days later from severe burns. Fearful of news of the accident getting out, like all their previous accidents at the time, the

Members of the Vostok program which included Yuri Gagarin, Gherman Titov and Pavel Popovich, seen here with Sergei Korolev. The cosmonaut Grigory Nelyubov is standing at the far back. (Roscosmos)

Soviets kept it secret, fearful that the West may think they were having problems, and it was to be decades before the truth about this incident, together with a number of others, came to light. There were others who were dismissed for various reasons including discipline, one of which was drunkenness, and so strict were the rules that one of the early cosmonauts, Grigory Nelyubov, had his picture erased from all the group photographs because of his erratic behaviour. Nelyubov turned to alcohol and later committed suicide, unable to face the disgrace of being thrown out of the elite cosmonaut corps.

CHAPTER THREE

The Development of the Manned Spacecraft

On 15 May 1960, the launch by a modified SS-6 rocket from Baikonur of *Vostok 1K*, the precursor to the manned space capsule, suddenly gave an air of urgency to the American space program. *Korabl-Sputnik 4*, the first of five 'manned' prototypes, Nos. 4, 5, 6, 9 and 10, was launched and placed in orbit at a perigee of 284km and an apogee of 514km. The primary role of this launch was that the main systems of the spacecraft/satellite could be monitored over a long period to ensure its continuing safe flight and control. The spacecraft's capsule contained the dummy of a man in a space suit, known to everyone as Ivan Ivanovich. It was also fitted with a television system and a number of scientific instruments. Experiments were also carried out using prerecorded messages and telemetry. After four days in orbit, the re-entry part of the spacecraft separated from the service module and fired the retro-rockets. Unfortunately the re-entry angle of the spacecraft was wrong and it burnt up as it entered the upper atmosphere. The remaining section of *Sputnik 4* remained in orbit until 15 November 1965, when it too burnt up on re-entry after spending 1,979 days in space. Part of *Sputnik 4* was found in the middle of a street in Manitowoc, Wisconsin, USA, but fortunately no one was hurt when the piece landed.

Then *Korabl-Sputnik 5* was launched on 19 August 1960, aboard a Vostok L carrier rocket, with a veritable zoo aboard. Together with Ivan Ivanovich were two small dogs by the name of Strelka and Belka, forty mice, two rats, and fifteen flasks of fruit flies. The purpose of the mission was to place a spacecraft into orbit with a cosmonaut, test the life support systems, study the effects of space and orbit on biological and botanical specimens and return safely to Earth. After one orbit, with an apogee of 20 miles (32.2 km) and a perigee of 190 miles (306 km), the capsule was separated and parachuted to Earth. All the animals were safely recovered, the first time living things had been recovered safely from space. A year after her ordeal in space, Strelka had a litter of puppies, one of which was given to President Kennedy's wife Jacqueline, as a goodwill gesture.

As the development of larger and larger rockets became more commonplace, it was realised that larger and more powerful engines were required. The R-7 motor was all right for the ICBMs and the early Sputniks, but to launch a rocket into space a much larger motor was required and this resulted in the creation by Mikhail Yangel of the

Vostok 1 spacecraft prior to being mated with its rocket. (RKS Energia)

Vostok rocket being transported to the launch site. (Roscosmos)

Vostok 1 launching from Baikonur. (Roscosmos)

R-16. The first test of this motor was a disaster. On 24 October 1960, as the countdown approached launch time, it was discovered that there was a leak of nitric acid from the base of the unmanned rocket at a rate of between 140–150 drops per minute. The technical management, aware of the pressure to launch the rocket, deemed it to be acceptable and sent an engineer from the chemical unit to keep it under control.

Experienced rocket scientists would have carefully drained off the fuel and then pumped non-flammable nitrogen into the tanks to purge them. Instead the Chief of Missile Deployment, Marshal Mitrofan Nedelin, sent in engineers to try and fix the leak. In the firing blockhouse, scientists should have reset all the electronic sequencers and then disarmed them, but they were ordered by Nedelin to delay the firing sequence and not cancel it. In charge of the test range, and ultimately responsible for safety of all the personnel there, was Major General Konstantin Gerchik. He could have activated safety procedures at any time that would have shut down the launch procedure, but Nedelin was his boss and there was no way that he was going to countermand his orders, such was the situation at the time. On the other hand, Marshal Nedelin was under considerable pressure from Premier Nikita Khrushchev to launch the rocket. Khrushchev was about to give a speech in the United Nations extolling the virtues of communism and the progress his country was making under his guidance. As the launch countdown continued, ground control sent a signal to activate the pyrotechnically operated membranes attached to the oxidizer lines on the second stage. The signal, because of a design flaw in the electrical circuits in the control panel, blew up the membranes on the fuel lines of the first stage. With these membranes destroyed, the fuel-laden rocket could not stay on the launch pad for more than two days – it was in fact a massive bomb waiting to explode. To complicate matters even more, minutes after the membranes blew up, explosive devices on the valves on one of the three engines in the first stage fired. The problems got no better when the electrical current distributor, which supplied power to the rocket, failed. Mikhail Yangel had argued constantly with Nedelin regarding the safety procedures, but as if to prove Yangel wrong, Nedelin ordered his staff to take their chairs and sit beside the rocket whilst it was being repaired. Yangel and his staff on the other hand went into the blast shelter, conscious of the danger of tampering with an unstable rocket. Somehow there was a confused signal given, and the wrong command was transmitted to the rocket's upper stage and the engine fired. The exhaust flame burnt through the fuel tank of the first stage and ignited the fuel inside. In the violent explosion that followed, about 190 engineers were killed (the exact figure is not known) and an unknown number were badly injured. Marshal Nedelin was amongst those killed in the explosion.

The Russians, well aware that the West would have known of the explosion, played it down, saying that Marshal Nedelin had been killed in an aircraft accident. For the next three years obituaries to the other technicians that died appeared spasmodically in the press, so as not to make the world aware of the disaster. It wasn't until 1989 that Russia allowed the truth to come out. Today there is a memorial to the scientists and technicians who died at the Tyuratam launch complex.

In an effort to draw attention away from the tragedy, another test flight of the Vostok spacecraft was carried out on 1 December 1960, when *Korabl-Sputnik 6* was launched. On board were two dogs, Pchelka and Mushka, together with a number of scientific instruments and a television system. After one day in orbit, re-entry was effected, but as the spacecraft entered the atmosphere, it was realised that the spacecraft's trajectory was incorrect due to a faulty firing of the re-entry rockets. Concerned that the spacecraft could fall outside of Soviet territory and could fall into foreign hands, explosive charges on board were activated, blowing it up and killing both dogs.

Marshal Mitrofan Nedelin. (Roscosmos)

The Russians became more ambitious and, on 4 February 1961, launched *Tyazhely- Sputnik 7* on a Molniya 8K78 carrier rocket into an Earth-parking orbit with the intention of later launching it towards Venus. But an ignition failure on the upper stage relegated it to orbiting the Earth as a satellite. Its large size led some western observers to think that it was a failed manned mission, but this was always flatly denied. On 26 February 1961 the spacecraft re-entered the Earth's atmosphere over Siberia, although originally it was thought that it had re-entered over the Pacific Ocean. This only came to light when a young boy, who was swimming in a lake in Siberia, hurt his foot on a piece of metal. On closer examination the sphere shaped piece of metal had some inscriptions etched into it. It was handed over to the local authorities who in turn sent it to Moscow to the Academy of Sciences. The sphere ended up on the desk of Sergei Korolev, who immediately recognised it as a pennant that had been installed in *Sputnik 7* from the Venus mission. Fortunately, because the sphere had been designed and built to withstand the hostile Venusian atmosphere, it had survived re-entry into the Earth's atmosphere.

The Russians took their space program one step further on 12 February 1961, when they launched *Tyazhely-Sputnik 8* on a mission to the planet Venus. The launch of *Sputnik 8*, which was the launching platform for the mission whilst in Earth orbit, sent *Venera 1* to Venus. This was a 'first' in many ways, inasmuch as it was the first spacecraft to be launched from an Earth orbit, the first with solar panels, a course-correction engine, 3-axis stabilization system and a parabolic telemetry antenna. Unfortunately three months after launch, a fault developed in one of the solar orientation sensors causing it to overheat, which in turn caused the telemetry to malfunction, and the spacecraft was lost. The spacecraft carried out a flyby of Venus, passing within 100,000km of the surface and a number of photographs were transmitted back before it was destroyed entering the planet's turbulent atmosphere.

On 9 March 1961, *Korabl-Sputnik 9* was launched and carried a dog by the name of Chernuska. In addition to the dog was a 'passenger' by the name of 'Ivan Ivanovich'. This was a life-size maquette complete with a spacesuit, which was made to look as lifelike as possible. A sign reading Maket (Dummy) was placed under his visor so that anyone finding him would not think that he was a dead cosmonaut or

Space dog Chernuska. (Roscosmos)

indeed an alien from another world. After a single orbit of the Earth, the spacecraft re-entered the atmosphere and the occupants were recovered safely. The mission was to test the onboard systems as well as carry out medical experiments on a living animal. This test was followed two weeks later on 25 March by *Korabl-Sputnik 10*. This spacecraft carried a dummy (Ivan Ivanovich again) and a small dog by the name of Zverdochka. After completing one orbit of the Earth, the spacecraft successfully re-entered and was recovered. The scene was now set for the first manned flight into space.

At 09:07 hours Moscow time, on 12 April 1961, came the breakthrough that was to change the face of space exploration. Russian cosmonaut Yuri Alekseyevich Gagarin was blasted into space from the launch site at Baikonur aboard *Vostok I*, call sign Kedr (Ceda), to become the first man in space and the first to orbit the Earth. Gagarin circled the Earth for 1 hour and 48 minutes at a height of more than 186 miles (300 kilometres) and at a speed of 17,400 miles per hour (28,000 kilometres).

Because the medical staff on the ground were unsure how the human body would react to weightlessness, during the flight Gagarin's controls were locked to prevent him taking over manual control, although a key to unlock the controls in an emergency was provided in a sealed envelope.

Dummy cosmonaut Ivan Ivanovich. The word Maket was written on the visor in case after a flight somebody reached him before the authorities and so would know that it was a dummy. (Roscosmos)

Yuri Gagarin in his spacecraft about to be launched. (Roscosmos)

Yuri Gagarin's capsule landed after completing the world's first manned space flight. (Roscosmos)

Sergei Korolev with Yuri Gagarin after his historic flight. (Roscosmos)

Above left: Yuri Gagarin with Nikita Kruschev at the Kremlin after his historic spaceflight. (Roscosmos)

Above right: Cosmonaut Gherman Stepanovich Titov, the second man to orbit the Earth. (RKK Energia)

The spacecraft was a variant of a new spy satellite, known as a Zenit. Engineers removed the cameras, installed a life support system and an ejection seat and redesigned the interior to accommodate a pilot. There were a number of problems with the spacecraft, the majority of which were minor but could have caused problems, but one of the main issues, that was discovered later, was that a key valve in the upper stage of the rocket had been assembled incorrectly at the factory. This could have caused a premature shutdown of the engine resulting in an unscheduled landing somewhere in Siberia, causing the recovery teams a major headache in finding the downed cosmonaut and his spacecraft. Fortunately it did not cause a problem, but one other did manifest itself and that was the survival pack containing supplies for three days, and which broke free from the ejection seat after Gagarin had ejected. In the event that he had landed away from civilization this could have been a major problem. There was one minor hiccup during re-entry and that was when the service module, after being separated from the capsule, was found to be still connected by a wire loom. The spacecraft started to gyrate violently but fortunately the wire loom burned through during re-entry. Gagarin ejected out of the spacecraft and parachuted to safety, landing near Engels Smelovka, Saratov.

Two young schoolgirls witnessed the landing of the spacecraft as a large ball that bounced once, leaving a large hole in the ground, before coming to rest. A farmer and his daughter saw Gagarin parachute to the ground, and described seeing a strange figure in a bright

Model of the SPIRAL spacecraft. (RKS Energia)

orange suit, wearing a large white helmet, walking towards them. Yuri Gagarin said, 'When they saw me in my space suit and the parachute dragging alongside as I walked, they started to back away in fear. I told them, don't be afraid, I am a Soviet like you, who has descended from space and I must find a telephone to call Moscow!' For a number of years the Russians denied this, saying that the cosmonaut had landed with his spacecraft, but finally admitted in 1978 that Gagarin had ejected prior to landing. Yuri Gagarin's life changed forever, he returned home to a hero's welcome and was fêted wherever he went in the world. His name was to be immortalized as being the first man to go into space.

In March 1964, Colonel Yuri Gagarin decided to extend his academic work to keep himself abreast of the space program and make himself available for the possibility of another space flight. A new course had been created, the Pilot-Engineer-Cosmonaut Diploma, a must for every budding cosmonaut. Gagarin chose as his thesis the design of a re-usable spacecraft, the dream of every space engineer. So radical were his designs that they were deemed to be top secret and no photographs of the drawings and designs were allowed.

Two weeks after Gagarin's spaceflight, on 6 August 1961, the Russians launched their second manned spacecraft, *Vostok 2*, call sign Oryel (Eagle). The cosmonaut was Major Gherman Stepanovich Titov, who had been Gagarin's back up on the first manned flight. Titov, at the age of twenty-five, was the youngest person to go into space and I believe still is. He carried out a 17-orbit, 25-hour flight, which was way beyond anything that the Americans had been able to even dream of doing, let alone achieve. During the flight Titov took over manual control of the spacecraft, carrying out several manoeuvres. He also suffered from prolonged space sickness during the mission that hampered some of the experiments he was assigned to carry out.

Cosmonaut Andrian Nikolayev. (Roscosmos)

During re-entry he experienced a similar problem to that of Gagarin's, the separation of the service module from the capsule being hampered by a wire loom. Fortunately the problem resolved itself by being burnt off as it came through the atmosphere and Titov ejected from the spacecraft successfully, landing near Krasny Kut, Saratov, Volga Federal District.

Gherman Titov never flew again; he was assigned to carry out testing and research on the Spiral space-plane, which many had declared to be a

dead-end project before it had even started. Gherman Titov retired from the Air Force as a Colonel-General and later became a very respected member of the Duma, the lower section of the Russian Parliament, until his death in 2000.

At the beginning of 1962, preparation was under way for another spaceflight, *Vostok 3*, call sign Sokol (Falcon), to be flown by cosmonaut Andrian Nikolayev. The flight was supposed to have taken place in March, but because an explosion during the launching of a Zenit-2 reconnaissance satellite had damaged one of the two Vostok launch pads, the launch was put back until 11 August. *Vostok 3*'s mission initially was to rendezvous with *Vostok 4*, which was to be launched the following day. The launch from Baikonur went according to plan and *Vostok 3* slid into orbit at an apogee of 218km above the Earth.

Andrian Nikolayev settled down to wait for fellow cosmonaut Pavel Popovich in *Vostok 4*, call sign Berkut (Golden Eagle), to be launched the following day, but in the meantime he had a number of experiments and scientific observations to be carried out. Further tests were carried out regarding the cosmonauts' ability to function under weightless conditions and carry out certain experiments. The main objective for this particular mission was to carry out a rendezvous, as close as possible, with another spacecraft. Neither spacecraft had any manoeuvrability capability, so the question of any link-up was not a viable one, but it did give the ground mission controllers practice in handling two spacecraft within such a close proximity of each other. After a quiet night, Pavel Popovich came within visual range of Andrian Nikolayev in *Vostok 4*, and the two spacecraft carried out their exercise. After completing his entire mission Nikolayev returned to Earth on 15 August, seven minutes after Popovich's spacecraft had touched down, both landing 200km apart near Karaganda, Kazakhstan. Both spacecraft had been scheduled for a four-day mission, but Popovich was brought down a day early because of a misunderstanding.

The misunderstanding came about because of the extreme paranoia that afflicted the Russian space scientists and hierarchy. Believing that everything that was said between the cosmonauts and the flight controllers were being monitored by the West, they created code words and expressions to pass sensitive messages back and forth between the cosmonauts and their controllers. In the case of Pavel Popovich, he had been instructed to say that if he felt ill due to space sickness, he was to say that he was 'observing thunderstorms'. As his spacecraft flew over the Gulf of Mexico, Popovich actually observed a violent thunderstorm and transmitted back to his controller at Baikonur, that he was watching the most violent of thunderstorms. He was recalled immediately and when back on Earth, it was discovered that there was nothing wrong with him and he had been more than willing to continue the flight.

An unmanned *Luna 4* spacecraft, built by the OKB-1 design bureau, was launched on 2 April 1963 aboard a Molniya-L 8K78/E6 rocket and was tasked to carry out a flyby mission of the Moon. After carrying out a low Earth orbit, the rocket stage of the spacecraft was fired up to send it on a course to the Moon. All was successful, but then during the trans-lunar coast, the astronavigation system failed to carry out a mid-course correction, causing the spacecraft to miss the moon by 5,282 miles (8,500 kilometres) and

ended up in a 55,825 miles (90,000 kilometres) by 496,476 miles (700,000 kilometre) orbit of the Earth. *Luna 4* eventually went into a heliocentric orbit of the Sun and was lost. It is thought that the mission was to attempt a soft landing on the moon, and a broadcast called '*Hitting the Moon*' was due to be made on the evening of 5 April had the mission been successful, but was cancelled when the failure of the mission was announced.

Romance suddenly entered the space race, when in June 1963, cosmonaut Andrian Nikolayev's girlfriend, 26-year-old Valentina Vladimirovna Tereshkova, joined the space program as a cosmonaut. Her first flight on 16 June 1963 was aboard *Vostok 6* and was programmed as a dual flight with cosmonaut Valery Fydorovich Bykovsky. *Vostok 5* was launched from Baikonur on 14 June 1963 on a Vostok 8K72K rocket, with the call sign Yastreb (Hawk), with cosmonaut Bykovsky on board. The purpose of the mission was to extend the information of the various space flight factors obtained on previous Vostok flights and to rendezvous with *Vostok 6*. Scheduled for an eight-day flight, the mission was hampered at the onset due to technical problems. Due to further technical problems after launch, *Vostok 5* was placed into a lower orbit than was planned and was noticeably going into a decaying orbit every time it orbited the Earth. Then, after only four days, Bykovsky contacted Mission Control, saying that he had severe problems with his waste management system and that conditions in the cramped capsule were becoming extremely uncomfortable and almost unbearable. The controllers realized that nothing more was to be gained from the flight and ordered it back to Earth the following day. As the spacecraft entered the atmosphere, the service module was jettisoned, but failure to release completely, causing the capsule to spin violently. Fortunately the heat from the re-entry burned off the offending retaining strap and the spacecraft resumed its normal re-entry mode.

Two days after the launch of *Vostok 5*, the last of the Vostok program, *Vostok 6*, was launched. Aboard the spacecraft, call sign Chayka (Seagull), was Russia's first woman cosmonaut, Valentina Tereshkova. The flight was intended to be a joint flight with *Vostok 5*, but because of the problems experienced by Bykovsky, it was decided to just use the flight to carry out research on the effect and comparisons of space flight on male and female organisms. It was suggested at one point, that putting a woman into space at the time was more of a propaganda exercise than a serious mission. This thought seems to have been endorsed, when Sergei Korolev, the director of space programs, was unhappy with Tereshkova's work rate and performance whilst in orbit and would not allow her to take over manual control of the spacecraft. Suffice to say that she never flew again. Tereshkova's flight lasted two days, in which she carried out thirty-one orbits of the Earth and carried out an extensive flight program.

One of Tereshkova's tasks was to take photographs of the horizon from space and these were later used in an attempt to identify aerosol layers within the Earth's atmosphere. Bykovsky, on the other hand, spent five days in space and completed eighty-one orbits of the Earth and numerous physical and psychological tests. Valentina Tereshkova later married fellow cosmonaut Andrian Nikolayev. Today Valentina Tereshkova is involved in Russian politics and is a respected member of the Duma and on several committees.

Valentina Tereshkova about to board *Vostok 6*. (Roscosmos)

Valentina Tereshkova taking a drink whilst in space. (Roscosmos)

A second unmanned probe was launched on 4 April 1964, when *Zond-1*, which was connected to a Tyazheliy-Sputnik atop a Molinya 8K78M rocket, lifted off the launch pad from Baikonur *en route* to Venus. Once in an Earth-parking orbit, the spacecraft, a Venera 3MV-1 that carried a spherical landing capsule, would be launched. It had originally been described as a deep space engineering test bed, but it was clearly designed to act as both a flyby and to launch a probe onto the surface. On reaching the planet's orbit the spherical capsule was to be launched towards the surface. The capsule contained a number of experiments, including a photometer, temperature and pressure gauges, a method of analyzing the atmosphere and of measuring the gamma rays of the surface rocks. However, not long after launch, one of the sensor windows developed a slow leak from a crack, causing the spacecraft to depressurize. This created a rarefied atmosphere within the spacecraft and when a command to turn on the radio system was activated inadvertently, it caused a corona discharge resulting in the electronics shorting out. Radio commands were then sent via the transmitter in the landing capsule. *Zond-1* went into a decaying orbit around the sun 100,000 kilometres from Venus after communications failed and was subsequently lost.

The Soviet Union, once the leader in the space exploration for so long, was now struggling to keep pace with the Americans. Premier Nikita Khrushchev, desperate to try and keep ahead of the Americans, ordered Korolev to work on his earlier design for

Vladimir Komarov with Konsantin Feokostitov (L) and Boris Yegorov (R). (Roscosmos)

a spacecraft that would take three cosmonauts. This meant removing the single ejector seat and squeezing three non-ejection seats into the already cramped confinements of the capsule. On 12 October 1964, they launched the first multi-crew spacecraft, *Voskhod I*, from their launch site at Baikonur on a Voskhod 11A57 rocket. Politics in the Soviet Union played a major part in the selection of cosmonauts and this became very apparent in the selection of the crew for *Voskhod 1*. The original crew consisted of Boris Volynov, Georgi Katys and Boris Yegorov, but three days before the launch, Georgi Katys was removed after the KGB had discovered that his father had been executed during Josef Stalin's 'Great Purge' of 1937. Boris Volynov was also replaced when rumours of his part-Jewish heritage came to light; he was later to be selected to fly on *Soyuz 5* and *Soyuz 21*. The replacement crew of Vladimir Komarov, Konstantin Feokostitov and Boris Yegorov had earlier in the year seen the successful launch of two unmanned Voskhod type spacecraft. The three cosmonauts, all of whom had different areas of expertise in engineering and science, had to diet and not wear pressure-suits in order to fit inside the spacecraft. The crew wore no space or pressure suits during the launch or whilst in flight; it was thought that the new spacecraft were now so well pressurized that they were unnecessary. This was to later prove fatal during the *Soyuz 11* mission. The crew were to conduct a number of physical, medical and engineering experiments during the mission. They were also tasked to carry out an in-depth medical-biological investigation program on the effects of weightlessness on the human body.

The first of these new spacecraft, *Voskhod 3KV*, with the three cosmonauts aboard, was launched from Baikonur on a Voskhod 11A57 rocket on 12 October 1964 and entered orbit at a perigee of 110 miles (178 km), reaching an apogee of 209 miles (336 km). Although the spacecraft was now designed to accommodate three cosmonauts, the designers had given very little thought to the re-siting of the instrument panels and left them in exactly the same place as they were in the original Vostok spacecraft. The crew completed a number of experiments before returning safely to Earth the following day. The Russian hierarchy claimed the mission to be a great success, but privately many in the space program regarded it as a 'political circus' due to the 'ridiculous' crew replacement saga. There was no escape tower fitted and had the launch malfunctioned there was no way that they could escape.

The second of the Zond unmanned spacecraft, *Tyazheily-Sputnik Zond 2*, was launched on 30 November 1964 from Baikonur on a Molniya T103 rocket. Once in a parking orbit, *Zond 2* was released, this time the target was the planet Mars. Very little is known about this mission except that it was designed to test the space-borne systems over a long duration in space and carry out scientific investigations during a flyby of the planet. One of the speculations surrounding this mission was that it was to make an unmanned landing on the surface of Mars. Most people, because of the higher energy trajectory, have accepted this, because the approach speed would have been greatly reduced and this would only make sense if the spacecraft intended to release a landing package, but there is no evidence to support this as the known description of the spacecraft doesn't support a landing module. In order to conserve power, the spacecraft communicated with its mission control for a couple of hours every few days. It soon became obvious that the spacecraft was operating on half power, after it was discovered that one of the solar panels had failed to open, which meant that a large number of the scheduled experiments would have to be cancelled.

In February 1965, a mid-course correction was carried out successfully, but by May, with communications becoming more and more erratic, ground control lost contact and it is thought that the spacecraft flew by Mars and headed out into the solar system.

CHAPTER FOUR

Baikonur

The Americans continued to make progress in their space program, and had gone from the single seat spacecraft (Mercury) to the two-seat (Gemini) and finally the three-seat spacecraft (Apollo) in stages over a period of years.

In July 1964, the Central Committee realised that because of the ongoing expansion of their space program, they would have to build a space centre that would encompass a factory for building the rockets, a launch site for launching them, accommodation for the scientists, engineers and workers and their families, together with all the additional facilities that would be required. The aim of the committee was still to beat the Americans to the Moon and the only way to do this was to build a massive launch complex and factory whatever the cost – and they approved it. The site chosen by the Russians for their space program was at Baikonur in the middle of the southern remote region of Kazakhstan in the Soviet Union. The area had been originally called Tyura but became known as Tyuratam by the nomads who traveled the region. Tam was the burial site of one of the sons of GenghKorolevis Khan, and was added to Tyura to include this. It had been built in 1945 as a launch site for missiles and later used for the early space rockets. The authorities changed the name to Baikonur in an effort to confuse Western Intelligence as to its location. Baikonur was in reality a small town some 250 miles just east of the Aral Sea and to the northeast of the Syr Darya River. The idea that it would confuse the west was a pointless exercise, because the moment the first rocket (an R-7 ICBM) was launched from the site, an American radar station in Turkey monitored the launch and then pinpointed the launch site. There was one other launch site, Kapustin Yar, situated sixty miles southeast of Vologograd, but this was used to launch the smaller type of rocket that was used for atmospheric research.

Made up of various sites, Baikonur covered an area 85 kilometres north to south and 125 kilometres east to west, but because of the break-up of the Soviet Union, it was ostensibly on 'foreign' soil (Kazakhstan), although it was, and still is, under the control of the Russian Federation.

In 1964, Chelomei's bureau NPO Mashinostroyeniya had designed Russia's first military space station, *Almaz*. This 20-ton spacecraft was to be used to carry out radar surveillance and photographic missions and would be serviced using a TKS (Transporting Korabl Snapzhenyia – Transport Logistics Spacecraft) spacecraft, a derivative of the Soyuz model. The original idea was to launch the *Almaz* space station into orbit, then once stabilized the crew and supplies would be launched in the TKS spacecraft. But Chelomei had other ideas and proposed that the *Almaz* was launched

The town of Baikonur. (Roscosmos)

together with the crew and the TKS spacecraft that was used to ferry crews and supplies. The space station would have a 'lifeboat' spacecraft attached in case of an emergency and would be used when the space station was finally discarded. Before any of these ideas could be formulated, political upheaval within the Soviet Union culminated with the removal of Nikita Khrushchev in October 1964, with the result that Chelomei fell from political favour and Korolev once again came to the fore. The battle for space supremacy between the two super-powers was still alive and Korolev and Chelomei were ordered to work together and design a new circumlunar program.

Two days before the Americans were to launch their first Gemini flight, on 23 March 1965 the second of the Voskhod missions, *Voskhod 2*, was launched on a Voskhod 11A57 rocket from Baikonur. This time there were only two cosmonauts on board, Alexi Leonov and Pavel Belyayev, and they were to carry out the first Russian space walk. The mission was fraught with problems from the first. After clambering into his space suit, Alexi Leonov carried out his space walk, but on his return he had the greatest difficulty in getting back inside the spacecraft due to the stiffness of his suit, and at one point it was thought that he wouldn't get back at all. Despite being almost unable to move his arms, he managed to reach the valve on his spacesuit and reduce the pressure sufficiently to give him mobility. It didn't end there. After he had got back in, it was found that the primary hatch would not seal properly. To compensate for this the environmental control system flooded the capsule with oxygen instead of the nitrogen-oxygen mixture that normally was in the spacecraft. This of course was

Baikonur taken by a U-2 spy-plane of the CIA.

a potential fire hazard as had been discovered in 1959, when cosmonaut Bondarenko died of burns suffered during training whilst in a similar environment.

After the space walk the spacecraft prepared for re-entry, but at the point of re-entry the primary retrorockets failed to fire so the spacecraft carried out another orbit of the Earth. At the point of re-entry, this time the retrorockets were fired manually. On entering the Earth's atmosphere the service module failed to release completely, causing the capsule to gyrate violently before the wires holding it burned through. The spacecraft landed deep inside a forest near Perm in the Ural Mountains, 2,000 kilometres away from the nearest rescue teams. The crew had to spend the night in their capsule surrounded by wolves, before being rescued the following day. Soviet doctors later reported that Leonov almost suffered heatstroke whilst in space, a condition that could have been fatal in that environment.

A lot of questions have been asked about this particular flight. Film and television pictures of the space walk were closely examined later, and a number of still unanswered questions were asked about why the two did not match up. Further missions that had been planned were scrapped and the Russian manned space program was put on hold.

Amongst the survival kits issued to the early cosmonauts was a 9mm semi-automatic pistol, which in the event of landing in the Siberian wilderness and having to fend off a 500lb bear or a pack of ravenous wolves, would be worse than useless. This was highlighted in 1965 when *Voskhod 2* landed 240 miles off course in the middle of a remote Siberian forest. The two cosmonauts, Alexi Leonov and Pavel Belyayev, well aware of their situation, realised that they would have to stay the night in their spacecraft. They were in an area where packs of wolves and wild bears roamed and realised that the only weapons they had were the 9mm pistols.

Fortunately they never encountered any wild animals, but Leonov voiced his concern for a purpose-designed weapon to be issued to all cosmonauts in their survival

Alexi Leonov carrying out the world's first space walk. (Roscosmos)

Voskhod 2 on the launch pad at Baikonur. (Roscosmos)

Sputnik 1. (Roscosmos)

kits. The result was the TOZ-82, also known as the TP-82. This was no ordinary pistol but one that was specially designed by Igor Aleksandrovich Skrylev and built for cosmonaut survival in the event of landing in the desolate wilderness that makes up a major part of Russia. This unique and unusual weapon had three barrels mounted in a triangular configuration. Two were smooth bore mounted side-by-side and fired 40-gauge shotgun cartridges, the third barrel, mounted beneath, was rifled and fired 5.45 x 39mm ammunition, the same as the AK-74 assault rifle. The pistol was loaded singly using the break mechanism, similar to that used in single-shot shotguns. The weapon was also capable of firing signal flares, making it a rather useful piece of survival kit. It was issued from 1986 to 2006, when, despite glowing endorsements from cosmonauts and American astronauts who had fired the weapon during training, it was withdrawn when it was discovered that the remaining ammunition for the gun had become unusable. In its place a standard 9mm semi-automatic pistol was issued, but before each mission it was discussed by space officials as to whether or not to issue one. The reasoning behind this, was it was thought that having weapons on a spacecraft or space station was fraught with problems and danger, especially after one crew member on one mission suffered a psychological episode. It is interesting to note that this was never considered when the TP-82 was issued on the earlier missions.

The Russians launched *Zond 3 Tyazheily-Sputnik* on 7 July 1965 on a Voskhod 11A57 rocket. Once in an Earth-parking orbit, *Zond 3* was launched towards the Moon and became the first of the Zond spacecraft to successfully complete a lunar flyby. It took a large number of photographs of the lunar surface and a number of the far side of

TP-82 cosmonaut survival weapon. (Roscosmos)

the Moon and transmitted them back to Earth. In total the photographs covered 19,000,000 square kilometres (7.3 million square miles) of the Moon's surface. After the flyby the spacecraft then went on to Mars, but although it missed the planet, it sent back a large number of images. Over the next few months it continued to send back images by radio until the vast distance (153 million miles) caused the radio reception to cease.

Zond 3. (Roscosmos)

With confidence growing in rocket development, the first of the Proton UR-500 8K82 rockets (*N4#1*) was launched on 16 July 1965 and placed in a low orbit for particle research. This was followed on 2 November by the launch of Proton UR-500 8K82 (*N4#2*), again for particle research.

The next of the Luna missions, *Luna 8*, was launched on a Molniya 8K78 rocket from Baikonur on 3 December 1965. This mission was to attempt a soft landing on the Moon, and all was going well, until just before the retro-rocket firing command was sent and the cushioning air bags were due to inflate, it was discovered that they had been pierced. The result was that it put the spacecraft into a spin and although some semblance of control was regained, it crashed into the lunar surface and was destroyed.

In January 1966, after a series of failed attempts, *Luna* 5, 7 and *8* all impacted on the moon, although *Luna 6* missed completely. *Luna 9* was launched on 31 January 1966 on a Molniya 8K78 rocket from Baikonur Cosmodrome and soft-landed on the surface of the Moon on 3 February.

The mission was to carry out a soft landing on the Moon and if successful, it would be the first to do so. The spacecraft itself consisted of two parts. The lander that achieved the soft landing had a spherical body which contained the radio and thermal control systems, the programming, batteries and scientific apparatus. On landing, four antennas mounted on the outside of the sphere would open automatically. The second part of the body consisted of the ascent stage, which contained the retro-rockets, four vernier rockets, fuel system, guidance and landing sensors, gyroscopes, the soft landing radar system and one small orientation engine. The landing was a complete success, but more importantly it proved that the lunar surface could support the weight of a spacecraft that would not sink into the lunar dust. It also gave the Americans some of the information they needed – a man could walk on the lunar surface. The landing announced to the rest of the world that the Russian efforts at this stage were more than a propaganda exercise. On touching down the spacecraft started transmitting images back to Earth after four sections that formed the top shell of *Luna 9* opened outward and stabilized the spacecraft on the lunar surface. The spring-controlled antennas then assumed their operating positions, and the television camera rotatable mirror system, which operated by revolving and tilting, began a photographic survey of the lunar surface. The first test image that was seen showed very poor contrast because the Sun was only about three degrees above the horizon, but was completed fifteen minutes later. Seven radio sessions, totalling eight hours and five minutes, were transmitted, as were three series of TV pictures. When assembled, the photographs provided four panoramic views of the nearby lunar surface, showing nearby rocks and a horizon over a kilometre away. The images also showed that the spacecraft was tilted slightly at about fifteen degrees and was close to the edge of a twenty-eight metre crater. At first the Russians were reluctant to show the images, but the Jodrell Bank Observatory in England, who had been monitoring the mission, noticed that the electronic signal being used to transmit the images was identical to that of the internationally agreed Radiofax system used

by newspapers for transmitting pictures. *The Daily Express* newspaper rushed a suitable receiver to the Observatory and pictures from *Luna 9* were decoded and published worldwide. *Luna 9* continued to transmit panoramic images back to Earth until the batteries ran out and the mission was ended on 6 February. In addition to the scientific instruments carried on the spacecraft, which included a gamma-ray spectrometer, a triaxial magnetometer and an instrument for measuring the infrared emissions from the Moon, it also carried a set of solid-state oscillators that had been programmed to play the musical notes of '*The Internationale*'. The intention was to broadcast it live to the 23rd Congress of the Communist Party of the Soviet Union, and during rehearsal on 3 April it played perfectly but the following day it was discovered that a 'note' was missing. Fortunately they had taped the first playing and so used that to the assembled Congress, claiming it was being broadcast directly from the Moon.

From their remarkable start, the Russian manned space program had just faded away, as the technology and finance required to put a man on the Moon seemed to have been put on hold. They even shelved plans to build a spaceplane called the Raketoplan that had been designed by the Russian designer and manufacturer Vladimir Chelomei, as a space interceptor.

Luna Lander 9. (Roscosmos)

The Russian space program was dealt a serious blow, when on the 14 January 1966, Sergei Korolev died unexpectedly during a botched haemorrhoid operation. Vasili Pavolich Mishin took his place as head of the TsKBEM (RKK Energia) also known as OKB-1. The OKB designation was given to all design bureaus and OKB-1 was specifically that of Korolev's. Known for his design work on the development of the N-1 rocket, Mishin was regarded as one of Russia's top rocket designers and engineers, but many thought he lacked the charisma of Korolev and the right political connections. Despite this Mishin got to work and co-operated with Chelomei in developing the N-1 rocket.

The proposal and designs were placed before the committee concerned in 1967, and approved in principle, with a launch date of 1970 to commemorate the 100th birthday of Vladimir Lenin. Chelomei regarded the date of the launch as completely unrealistic and argued that he would need possibly another two to three years beyond that date.

With the American Apollo program well underway, the Russian leaders became concerned that they were falling way behind, and decided to up the stakes by ordering Mishin's bureau Energiya NPO to build a space station, using the design of the *Almaz* station as a basis. As if to rub salt into the wound, Chelomei was ordered to transfer all the drawings of the *Almaz* to Mishin's design bureau. Within one year two flight-rated *Almaz OPS* (as it was now known) spaceframes had been completed by Chelomei's bureau (OKB-52) along with eight ground test articles. All the frames and test articles were then passed to the Mishin bureau for implementation to their 17K (DOS) (Dolgovremennaya Orbitainaya Stanzlya – Long Duration Orbital Station) project. In addition Chelomei was ordered to use Korolev's design of the Soyuz spacecraft instead of the TKS during test flights; the first of these being *Salyut-1*.

Slowly but surely Chelomei removed all of Korolev's designs of the *Almaz* space station from the finished article and the TKS re-supply craft replaced the Soyuz-R. The *Almaz* project had been created specifically to carry out a piloted observation of American military installations from a low-Earth orbit. It comprised of two components, an Orbital Piloted Station (OPS) and a Transport Supply Ship (TKS). The crews for the space station were drawn from all the services: Army, Navy and Air Force. Each would be trained in their area of expertise, for example, missile men, submariners and pilots.

As with all these types of projects problems surrounded manufacture. Nothing on this scale had been tried before. In theory there was no problem, but in practice delays followed by more delays caused the military to re-think the whole project.

A third Voskhod flight, *Voskhod 3*, took place on 22 February 1966. Also known as bio-satellite *Kosmos 110*, in addition to a number of scientific instruments carried aboard, were two dogs, Verterok and Ugolyok. The spacecraft was in orbit for 23 days and on 16 March landed safely. Both dogs were in good health and lived out their remaining lives as the pets of members of the ground crew. Their endurance record for animals in space still stands to this day and was only surpassed by humans on the *Skylab 2* mission.

Whilst everything seemed to be concentrated on a lunar program, research was still being carried out on the development of the rocket and on 24 March 1966 Proton

UR-500 *8K82* was launched. Unfortunately the second stage engine failed on separation from the first stage, causing the two stages to collide with a catastrophic result.

Despite the lunar mission setbacks, another unmanned space mission took place when *Luna 11* was launched on 24 August 1966. The spacecraft sent back a total of 137 radio transmissions and completed 277 orbits of the Moon. A problem arose when the spacecraft lost its orientation when one of the thrusters became jammed, causing the television cameras on board to face away from the surface of the moon; consequently no usable photographs were returned. The mission ended when the batteries failed and the spacecraft was lost.

Two months later, on 22 October, *Luna 12* was launched from Baikonur on a Molniya-M 8K78M rocket. It was then launched from an Earth-orbiting platform to the Moon to complete the mission that *Luna 11* had failed to do. This time the television cameras operated perfectly and numerous photographs of the surface of the Moon were sent back to Earth from lunar orbit; the exact number is not known. Two of the main areas covered by the cameras were the Sea of Rains and the Aristarchus crater. After over two months of communication with the spacecraft, contact was finally lost. Because of the interception of the images transmitted by *Luna 9* by the Jodrell Bank Observatory, the transmitting signal was changed.

On 14 December a Soyuz rocket was launched to carry out a rendezvous with *Kosmos 133*, but it exploded on the launch pad. At the command of lift-off, the core stage ignited, but not the strap-on rockets. An immediate shutdown command was issued and the pad crews moved to drain the fuel and put the service tower back. Suddenly, as this was being done, the launch escape system fired causing the third stage tanks to overheat and explode. One member of the ground crew was killed and a number of others injured, putting the program on hold.

Russia continued with their lunar program when they launched *Luna 13* on 21 December 1966, only this time the mission was to soft-land the spacecraft on the moon's surface. Launched from Baikonur on a Molniya-M 8K78M rocket, the spacecraft landed on the lunar surface three days later in the Ocean of Storms region, and minutes later the first transmissions were being sent back to Earth. The cameras operated in a panorama vista, taking 100 minutes to send back a complete panorama of the silent desolate world. In addition to the cameras, the spacecraft was equipped with a soil measuring penetrometer, and a radiation densitometer to measure the cosmic rays that bombarded the surface of the Moon. These rays, when measured, proved to be less hazardous for humans than was as first thought. One of the two cameras on board failed, but by this time the batteries were almost drained and after a very successful mission, contact was finally lost on 28 December.

Another unmanned mission, *Kosmos 140*, was launched on 7 February 1967 and, after orbiting the Earth for four days, was successfully brought back to land in Kazakhstan.

Two more Voskhod missions were planned, but both were cancelled so that the designers and engineers could concentrate all their efforts on getting a manned spacecraft to the moon. In the meantime tests were still being carried out on the Proton rocket, and 10 March 1967 saw the launch of the Proton-K UR-500K *8K82K* (later *Proton-K*) on

Vladimir Mikhailovich Komarov. (Roscosmos)

a successful short test flight. One month later a second test flight, *Proton-K/D*, ended in disaster when the fourth stage separated prematurely. Then disaster struck the Russian space program when on 23 April 1967, *Soyuz (Union) 1* was launched with Cosmonaut Vladimir Mikhailovich Komarov on board. The Russian Premier Leonid Brezhnev made it quite clear that he wanted to celebrate the fiftieth anniversary of the founding of the Soviet Union with something spectacular from their space program and 'suggested' that a rendezvous in space between two Soviet spacecraft would be perfect. The idea was that the first spacecraft (*Soyuz 1*), flown by Vladimir Komarov and another cosmonaut, would launch first, followed closely by a second spacecraft (*Soyuz 2*), flown by Yuri Gagarin, also with a second cosmonaut. The two spacecraft would then dock whilst orbiting the Earth. The plan was that Komarov would exit *Soyuz 1* and make his way to *Soyuz 2*, then exchange places with Yuri Gagarin, who would then take Komarov's place in *Soyuz 1*. With the exchange completed the two spacecraft would then return to Earth and a hero's welcome.

However it was decided to launch *Soyuz 1* with just Vladimir Komarov on board with Yuri Gagarin as the back-up cosmonaut. Komarov and Gagarin were very close friends and both knew that the spacecraft was not safe to go into space. There had been reservations about the flight some months before, detailing some 200 defects in the new Soyuz spacecraft. Komarov knew of these, and expressed a reluctance to go on the flight, but realised that his friend Yuri Gagarin, being the back-up pilot, would have to take his place and he didn't want his friend to risk his life if anything went wrong. Just before the flight, Gagarin and a number of the other cosmonauts prepared a document detailing the 200 defects, requesting that the mission be aborted. Gagarin gave it to a close friend in the KGB by the name of Venyamin Russayev, and asked him to send it up the chain of command to Premier Leonid Brezhnev, but it was never sent. The Russian hierarchy wanted desperately to have something special to celebrate the fiftieth anniversary of the Russian Revolution, so the document never left KGB headquarters. There were strong mutterings within the space fraternity regarding the mission, all saying that it should be postponed because the spacecraft wasn't safe. It was even suggested that Vladimir Komarov should refuse to fly on the mission, but no one dared to say a thing to Brezhnev, knowing full well who ever did would suffer the wrath of the Premier.

The launch on 23 April 1967 of *Soyuz 1* on a Soyuz 11A511 rocket went as planned, but just after orbital insertion things started to go wrong. First one of the solar panels failed to deploy, wrapping itself around the service module, reducing the spacecraft's electrical power by fifty percent. Then the spacecraft started to rotate on its own axis and got progressively worse as Komarov tried to correct it, preventing the spacecraft from manoeuvring. Ground control knew they had serious problems and the decision was made to bring the spacecraft back to Earth, The tragedy was that Komarov knew the problems were very serious and was well aware that his chances of survival were slim. The seriousness of the situation was made even more apparent, when Premier Alexsei Kosygin, who was sharing power with Leonid Brezhnev at the time, himself called Komarov personally on a videophone link and who was heard to be crying. Komarov's wife also came on the link-up and Komarov dictated to her how she was to take care of the children and his affairs. On the ground, because of the ongoing problems, it was decided to postpone the launch of *Soyuz 2* and cancel the whole program. Finally Komarov managed to get the spacecraft under control and fired the retro-rockets manually. Re-entry was made safely, and at the correct altitude the drag chute deployed. Then, because of a faulty pressure sensor, the main chute did not deploy, so Komarov released his reserve chute manually, but the parachute straps got twisted with the drogue chute. U.S. monitoring stations in Turkey picked up the last few moments of Komarov's life as he ranted and raged at the men who built the faulty spacecraft. The spacecraft crashed into the ground with the force of a three-ton meteorite, near the town of Orenberg, killing Komarov instantly. The cabin exploded on impact and when Soviet Air Force recovery teams arrived all they found was burning metal, the rim of the top of Soyuz being the only hardware they could identify.

Vladimir Komarov was given a state funeral and his ashes were buried in the Kremlin wall with great pomp and ceremony. This put the Russian space program on hold until the cause of the incident was found, and all their other spacecraft could be modified to prevent it happening again. It was rumored that the designer *of Soyuz 1*'s parachutes and release equipment, Pavel Tkachev, was imprisoned for two years after Komarov's death. Because it was almost impossible to pinpoint which individuals were responsible for the 200 defects known to have been in the spacecraft, Pavel Tkachev was the obvious and easiest one to hold responsible and so became the scapegoat.

The first automatic rendezvous and docking of two unmanned spacecraft took place on 30 October 1967, when a docking target, known as *Kosmos 186*, docked with *Kosmos 188* at the second attempt. Initially Sergei Korolev had advocated the use of the automatic docking system and had faced fierce opposition, but after bitter arguments his idea was accepted. This was proved to be right and the successful mission was marred only when *Kosmos 188* misaligned just before re-entry. This was because of the failure of one of the on-board ion sensors that caused it to deviate from its landing course and the vehicle had to be destroyed using the on-board self-destruct system. *Kosmos 186* also suffered from a similar problem, but mission control managed to regain control and guide the spacecraft to a soft landing at the pre-arranged landing zone.

Remains of *Soyuz 1* spacecraft in which Komarov died. (Roscosmos)

Tests continued on the Proton rockets, and two more test flights, one in September and one in November, failed after problems in the launch sequences. Then on 2 March 1968, the unmanned *Zond 4* mission took place to test the space-worthiness of the new spacecraft, which was in fact almost another version of the Soyuz-7K-L1 spacecraft. It was equipped with new systems and equipment, and was the first Russian spacecraft to have a computer on board. Launched on a Proton D-1e rocket, *Zond 4* reached a height of 300,000 kilometres. The spacecraft was then set up for re-entry, but ground control had made the angle of descent too steep and it hurtled through the atmosphere over West Africa. It was realised that the spacecraft was not going to come down on Russian soil, so over the Gulf of Guinea they activated the destruct mechanism and the unmanned spacecraft exploded.

Another human tragedy beset the Russian people on 27 March 1968, with the death of Yuri Gagarin in a plane crash. Together with his instructor Colonel Vladimir Serugin, Yuri Gagarin took to the air from Chkalovsky, near Star City, in a MiG-15UTI jet fighter on a familiarization flight. Gagarin was looking to qualify to fly the MiG-17 fighter, his flying having gone-by-the-board after his famous spaceflight. Much speculation has been made regarding the crash, even to the point of almost accusing the Russian government of arranging it to rid themselves of Gagarin. It was even said that Gagarin's world-wide popularity was causing a great deal of resentment amongst certain members of the Politburo and that his crash was no accident. His heated face to face meeting with Leonid Brezhnev over the death of his close friend Vladimir Komarov, caused many to think he may have overstepped the mark and it was rumored that at one point he threw a glass of champagne over Breznev. There have been many questions and conspiracy theories about his death, but the consensus of opinion, after an in-depth

Last known photograph of Yuri Gagarin. Said to have been taken just before his fateful jet flight. (Gromov Institute)

investigation, was that his aircraft went into an uncontrollable spin and crashed after a Sukhoi SU-15, flying below Gagarin's MiG-15, created a strong turbulence as it passed through the sound barrier. This account was supported by fellow cosmonaut Alexi Leonov, who was in the vicinity at the time, and said that he heard two 'booms'; the first was a supersonic one and the other, some thirty seconds later, was probably Gagarin's MiG-15 hitting the ground. An investigation by the KGB however came up with another version. They said that Gagarin had been given the wrong weather forecast, saying that the cloud level was at an acceptable height, when in fact it was at almost ground level. No one will ever know the truth as the only two people who really knew what happened died in the crash. The cosmonaut/astronaut world were left to mourn the passing of the world's first man to go into space.

The last in the series of lunar missions took place on 7 April 1968, when *Luna 14* was launched from Baikonur on a Molniya-M 8K78M rocket. It entered lunar orbit three days later on 10 April and carried out communication experiments in preparation for a possible lunar landing. Controllers were also able to track the spacecraft, which enabled them to map and gain information on terrain anomalies that may affect any landings. The mission lasted seventy-five days before the power failed and all communication was lost. It is assumed that *Luna 14* then crashed onto the surface of the Moon.

The first completely successful unmanned Soyuz spacecraft to rendezvous and dock with another Soyuz spacecraft, using ion sensors to detect the direction of motion, was carried out on 14 April when *Kosmos 212* was launched from Baikonur on a Soyuz 11A511 rocket. *Kosmos 213* was launched the following day on a Soyuz 11A511 rocket and both orbited the Earth before making contact with each other. There was concern at first when ground control lost contact with the spacecraft, but then another tracking station confirmed that they had received telemetry signals to indicate a successful docking. After almost four hours docked together, the two spacecraft separated. *Kosmos 212* landed without any problem in the Kazakhstan area on 19 April. The following day *Kosmos 213* touched down in the Kazakhstan area in the middle of a violent storm, everything working perfectly until it came to jettisoning the parachute, which didn't activate. This caused the capsule to be dragged for several kilometres before finally coming to a halt.

It was argued that if the spacecraft had been manned then the parachute would have been manually jettisoned immediately on touchdown. But the death of Vladimir Komarov some weeks earlier was still fresh in the minds of everyone. The way was now clear for a manned mission to take place, and the cosmonaut Georgi Beregovi always claimed that all flights should be manned and that Komarov's death in April 1967 was nothing more than to be expected when testing untried and new forms of aviation.

A second tragic accident happened on the launch site at Baikonur on 15 July 1968, when three pad engineers were killed when the fourth stage oxidizer tank of the L-1 rocket they were carrying out tests on over-pressurized and exploded.

The final test of the Soyuz spacecraft prior to the launch of *Soyuz 2*, was the launch of *Kosmos 238* on 28 August 1968. This modified version tested the orbital manoeuvring system, re-entry, descent and landing systems successfully.

Another flyby mission to the Moon took place on 14 September 1968, when *Zond 5* lifted off the launch pad at Baikonur. The mission was launched using the UR-500K Proton K/D rocket and placed in an Earth-parking orbit. Similar in design to *Zond 4*,

Proton K rocket with a Zond spacecraft attached, being moved to the launch pad. (RKK Energia)

the *Zond 5* capsule was cylindrical, 4.5 metres in length and 2.72 metres in diameter, with two solar panels, which opened once in Earth orbit, giving the capsule a total diameter of 9 metres.

A problem with the stellar attitude control system en-route to the Moon caused it to initially become unusable, but backup sensors enabled the spacecraft to be kept under control. On board the spacecraft was a biological payload consisting of plants, seeds, flies, meal worms and turtles. In the pilot's seat was a 175-cm tall mannequin, which contained scientific instruments that measured radiation. This mission had been scheduled to be a manned flight to the Moon, but because of the failure of *Zond 1A* and *Zond 1B* the authorities decided to send an unmanned version of the Soyuz 7K-L1 spacecraft. The spacecraft arrived in lunar orbit three days later where it took some of the highest quality photographs of the lunar surface ever seen. The mission up to this point had been a relative success, but on its return to Earth one of the backup attitude control sensors failed, making it impossible for the controllers to make a planned guided re-entry. It was decided to make a direct ballistic re-entry using the Earth's atmosphere as a means of braking the spacecraft aerodynamically. At 4.5 miles (7 kilometres), the parachutes were deployed and the capsule splashed down in the Indian Ocean. The spacecraft was recovered by one of the two Russian recovery ships, the *Borovichy* and the *Vasilly Golovnin*, that were patrolling in the area.

Zond 5 in the Indian Ocean awaiting recovery. (Roscosmos)

The tragic death of Vladimir Komarov had set the Russian manned spaceflight program back, and it wasn't until 25 October 1968 that the next flight of a Soyuz spacecraft, *Soyuz 2*, was launched from Baikonur on a Soyuz-2 1a Rocket. This was an unmanned spacecraft, although there had been rumors that it was a manned flight, with cosmonaut Ivan Istochnikov at the controls and accompanied by a dog by the name of Kloka; this was completely untrue as there is no evidence to support this. In fact there is no record of any cosmonaut with that name! The following day, 26 October, a second launch was carried out with *Soyuz 3*, this time it was piloted by a genuine cosmonaut – Georgi Beregovoi. Launched from Baikonur on a Soyuz rocket, it had been intended to carry out a rendezvous between *Soyuz 2* and *3*, at an apogee of 205km, but despite numerous attempts, the closest Beregovoi got was within one metre and was unable to dock with the unmanned spacecraft. The blame for this was placed squarely on Beregovoi, who, it is said, failed to notice initially that *Soyuz 2* was upside down and repeatedly put his spacecraft into an orientation that made docking impossible. He used almost all of his 80kg of fuel in the first attempt, making a second attempt very brief and any more attempts out of the question. He could not use the remainder of the fuel because between 8–10kg was required to orientate the spacecraft for retrofire and to effect re-entry. After completing eighty-one orbits of the Earth for the next two days in which he made a number of meteorological and topographical observations, the spacecraft re-entered the Earth's atmosphere and landed safely on 30 October in a blizzard just outside Karaganda, Kazahkstan. Despite the docking incident, the mission was deemed to be a great success and Georgi Beregovoi was promoted to Major General and made Director of the National Center for Cosmonaut Training at Star City.

Zond 6 was launched on the 10 November 1968 from Baikonur on a Proton-K11 rocket, for another unmanned flyby of the Moon. Although another test flight for the proposed manned circumlunar flight, the spacecraft carried a number of scientific probes including micrometeoroid detectors, a cosmic ray detector, a biological payload as well as the usual photographic equipment. The flight was a complete success until problems arose on the return leg. A faulty rubber gasket in the unmanned crew section caused the spacecraft to slowly depressurize, killing all the living specimens on board; this of course would have been fatal had there been a human crew aboard. As the spacecraft dipped into the atmosphere over Antarctica, the parachutes deployed too early, causing the main canopy to rip and the capsule to subsequently crash in Kazakhstan. Fortunately the self-destruct system failed to activate and the on-board camera was recovered.

The reason given why the parachutes deployed early was put down to the vacuum inside the partially depressurized capsule that had been created by the leak, causing static electricity to build up in the spacecraft's electronics. An erroneous signal, caused by a coronal discharge from the Sun, was sent to the gamma altimeter making it think that landing was imminent. This in turn triggered the soft landing rockets to fire, which was followed by the deployment of the drogue chutes and then the main chute whilst the spacecraft was still 5,300 metres above the Earth.

Because of the problems experienced with *Zond 6*, the launch of *Zond 7* was cancelled until August 1969.

On 21 December 1968, the Americans launched *Apollo 8*, with astronauts Frank Borman, Jim Lovell and William Anders aboard. The spacecraft entered lunar orbit on 24 December and carried out a series of photographs of both the dark side and the nearside for potential landing sites. A number of scientific experiments were also carried out. The Americans, despite the strenuous efforts of the Russians to be the first, had achieved the first manned orbit of the Moon.

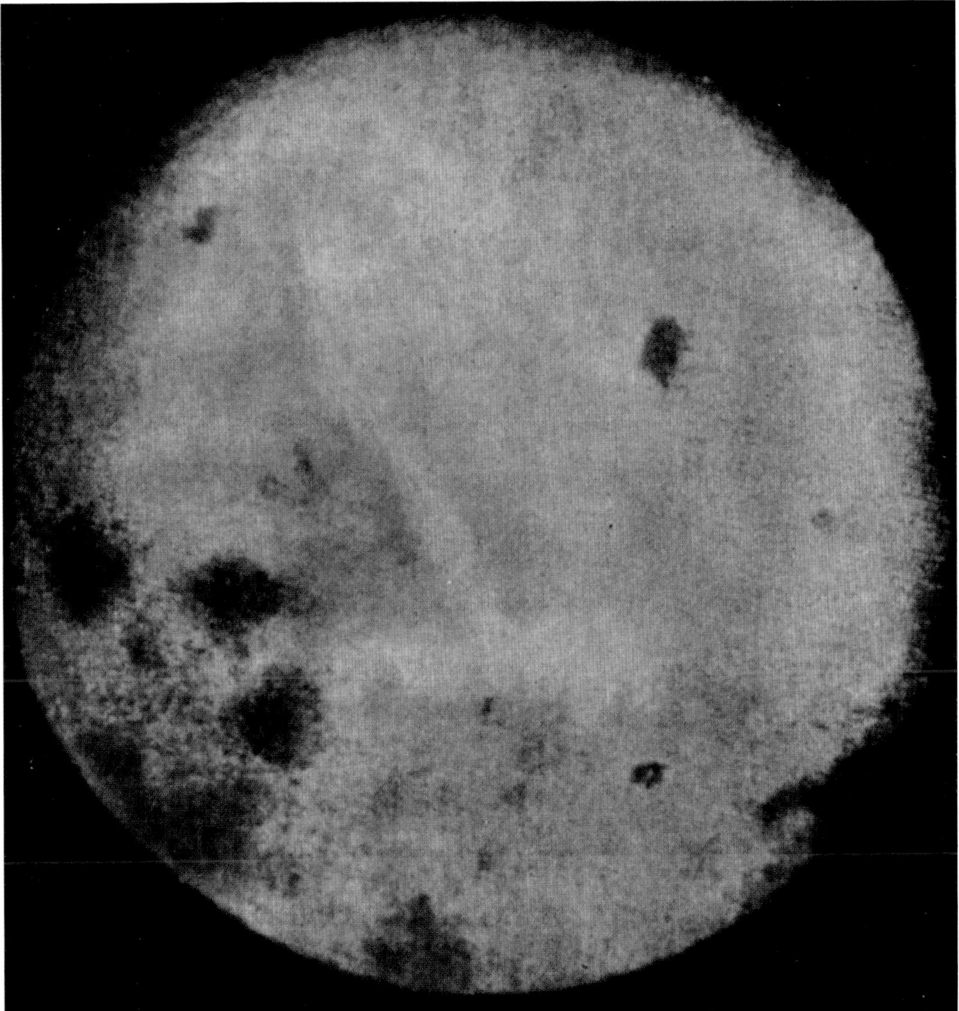

Photograph of the dark side of the Moon. (NASA)

CHAPTER FIVE

Venus and Lunar Missions

Russia's space exploration program continued with the launch on 5 January 1969 of *Venera 5* on a mission to Venus to obtain data about the planet's atmosphere. The spacecraft reached the planet on 15 May and jettisoned a capsule containing scientific instruments. The capsule descended through the Venusian atmosphere on a parachute that was designed to take it to the surface in about fifty-three minutes. Data was being transmitted back to Earth almost as soon as the capsule started its descent. For the next fifty-three minutes the spacecraft transmitted data back to Earth, but then ceased soon after the spacecraft touched down, due to the planet's immense pressure of 2,610 kilopascals (26.1 bar) and the 320°C (608ºF) degree temperature. The mission was regarded as having been successful, despite the loss of the spacecraft. *Venera 6* was launched five days later and returned similar results.

Although the *Soyuz 3* mission was deemed to have been a partial failure because of the problem with docking, it had been brought safely back. With this in mind they launched *Soyuz 4* on 14 January 1969 from Baikonur on a Soyuz rocket, with cosmonaut Lieutenant-General Vladimir Aleksandrovich Shatalov on board. Then *Soyuz 5* was launched the following day, with cosmonauts Colonel Yevegeniy Vasil'yevich Khrunov, Colonel Boris Valentinovich Volynov and civilian engineer Aleksey Stanislovovich Yeliseyev aboard. The two spacecraft rendezvoused and docked during the eighteenth orbit and prepared to transfer Khrunov and Yeliseyev from *Soyuz 5* to *Soyuz 4*, the first time that crews had transferred from one spacecraft to another whilst in space. This was an important breakthrough, because one of the problems that had always concerned both cosmonauts and astronauts was, that if they were to have a problem whilst in space, how would they be rescued in the event of them being unable to return? On the thirty-fifth orbit, Khrunov exited *Soyuz 5* first and started to make his way over towards *Soyuz 4*. As he did so, his lines became tangled and it took some time to untangle himself. During this, Yeliseyev, who was watching his friend intensely trying to untangle himself, had forgotten to mount the video camera to record the historic event. With the lines untangled, Khrunov entered *Soyuz 4* and waited for Yeliseyev to join him some minutes later, which he did without incident. The two spacecraft separated and returned to Earth. *Soyuz 4* now with three cosmonauts aboard and *Soyuz 5* with just one. The two cosmonauts who joined Shatalov brought with them some letters and a newspaper, thus delivering the first 'space mail' and proving that the exchange had taken place.

Above: The surface of Venus

Below: Venera 4 spacecraft

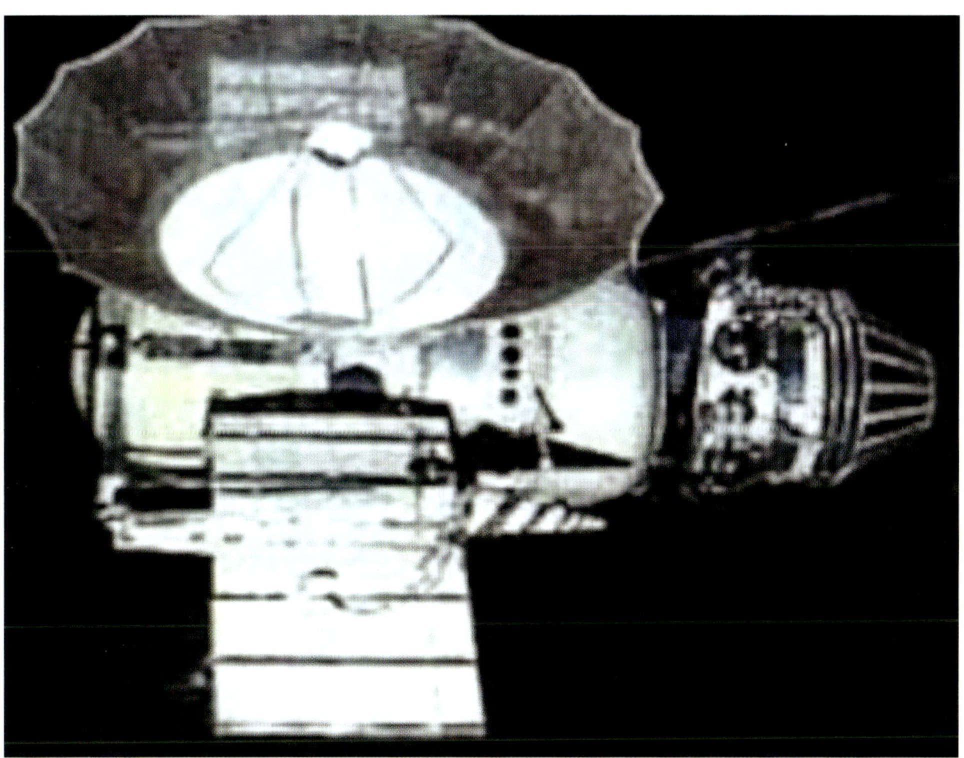

Soyuz 4 returned and landed south-west of Karaganda, Kazakhstan, in a temperature of -30 degrees C and in 80cm of snow without incident, but *Soyuz 5* on the other hand was involved in one of the most dramatic re-entries in the history of space travel up to then. The 'capcom' (capsule communicator) Belyayev asked Volynov to set the spacecraft for a manual retrofire prior to re-entry, but Volynov had already set the spacecraft up for automatic retrofire. *Soyuz 5* started its re-entry and fired its retro-rockets to separate from the service module, but the module failed to part from the capsule. The two sections of the spacecraft hurtled towards Earth and as the atmosphere increased, the spacecraft went into the worst aerodynamic position and approached nose first. This meant that the spacecraft had no heatshield to protect it and the light metal entry hatch at the front was in danger of melting. Fortunately the friction melted the fragile struts holding the service module to the capsule, and the descent module righted itself before the hatch melted. At the pre-programmed altitude the soft-landing rockets, which were supposed to fire and slow the spacecraft down, failed to ignite, but the drogue chutes did deploy followed by the main chute. The landing itself was extremely hard and Volynov broke some of his teeth in the impact.

It was reported later that Vasili Mishin, who was in charge of the program, was not available to assist in the making of the key decisions regarding the re-entry, because he was celebrating with the crew of *Soyuz 4* and was still drunk the following morning when Leonid Brezhnev called demanding to know what was going on.

The four cosmonauts were invited to Moscow to be awarded for their achievement at a ceremony in the Kremlin. Premier Brezhnev was to attend and the two crews were waiting on a podium in Red Square waiting for the premier's cavalcade to drive into Red Square, when, as one of the two limousines swept into the square, Viktor Ivanovich Ilyin, a deserter from the Soviet Army and dressed as a policeman, stepped forward from the crowd with two automatic pistols in his hands and fired eight shots at point blank range into the second car in an attempt to assassinate the Premier. However in the second car were cosmonauts Alexi Leonov, Andriyan Nikolayev, Valentina Tereshkova and Georgy Beregovoy. Fortunately none of them were hit, but Brehznev's car raced past the podium and the waiting cosmonauts, and into the Kremlin itself. A police motorcyclist ran the would-be assassin down and he was arrested. The presentation was held later inside the Kremlin. Viktor Ivanovich Ilyin was interrogated by the KGB and declared insane. He was then placed into a secure psychiatric hospital and kept in solitary confinement until his release in 1988.

The first launch of the N-1 *L-3* took place on 21 February 1969 and was in trouble immediately it lifted off the launch pad. High oscillations in the gas generator of number 2 engine caused some of the components to tear loose from their mountings. This resulted in a leak of propellants, which then ignited in the tail compartment. The engine monitoring system detected the fire, but then gave an erroneous signal to shut down all the engines just sixty-eight seconds into the flight. Consequently the range safety officer ordered the rocket to be destroyed. Any thought of beating the Americans to the Moon were dashed: the Russian space program had fallen woefully behind. It has to be remembered that over twice as many people were working on the American space program as on the Russian space program, and that they had a great deal more money and freedom.

Another N-1 rocket was launched on 3 July 1969, the N-1 *L-5*. Within seconds of it being launched the rocket exploded, destroying the launch pad. It was discovered later that the oxidizer pump on engine No.8 had ingested a piece of foreign matter, causing it to shut down. Twelve seconds later the remaining engines, with the exception of No.18, also shut down. The whole rocket then started to slide back down but No.18 engine caused it to tilt as it did so. The escape tower fired taking the upper shroud and re-entry capsule with it, whilst the rocket crashed onto the launch pad and exploded. Thirty-five kilometres away the town of Leninsk rocked and the townspeople suddenly saw a huge ball of fire light up the sky. Despite this setback the fifteenth Luna mission took place on 13 July 1969, when *Luna 15* was launched from Baikonur on a Proton K/D rocket in an attempt to carry out a soft landing of the surface of the Moon. It had been intended for the spacecraft to carry out a second sample-collecting mission in the Mare Crisium region, but it crashed onto the lunar surface and was destroyed.

When *Apollo 11* landed on the Moon in July 1969, and Neil Armstrong and Edwin 'Buzz' Aldrin stepped down onto the surface, the race for the first manned landing on the Moon was over. Having failed to beat the Americans to the Moon, the Soviet Union's aim now was to be the first to put a space station into orbit around the Earth. Work on the project started in earnest, focusing their efforts on carrying out rendezvous experiments.

On 7 August 1969, the unmanned *Zond 7 L1* spacecraft was launched on a Proton K/D rocket, with a life-size dummy, known as FM-2, aboard. The dummy was made of plastic and was fitted with internal and external instruments that would measure the dosage of radiation

The test dummy Ivan Ivanovich. The likeness to Yuri Gagarin is very apparent. (Roscosmos)

that a cosmonaut would be subjected to during an orbit of the Earth or the Moon. It also had the face similar to that of Yuri Gagarin given in homage to the late cosmonaut. The mission was a complete success and the spacecraft returned to Earth on 14 August, landing south of Kustanai, Kazakhstan.

Meetings in Moscow between the chief designers, engineers and the various committees that oversaw the space program, were becoming more and more heated. Mstislav Keldysh launched a full-scale attack on the continuing efforts to put a man on the Moon. The Moon, in his opinion, could be explored just as easily with robots and probes far less fraught with problems and more economically than a manned mission.

On 11 October 1969, *Soyuz 6* with cosmonauts Georgi Stepanovich Shonin and Valeri Nikolayevich Kubasov aboard was launched. The following day *Soyuz 7* was launched with cosmonauts Vladislav Nikolayevich Volkov, Anatoli Valis'yevich Filipchenko and Viktor Vasil'yevich Gorbatko aboard. They carried out rendezvous manoeuvres whilst in orbit round the Earth and then were joined by *Soyuz 8*, which was launched on 13 October 1969 with cosmonauts Vladimir Aleksandrovich Shatlov and Aleksey Stanislovovich Yelisyev aboard. The three spacecraft carried out a variety of experiments all centred around rendezvous techniques, but because of a communications failure in the radio location system, *Soyuz 7* and *Soyuz 8* never got closer than 1,600 feet. The whole exercise was to have been filmed by *Soyuz 6*, but it never got close enough to do so.

On 11 April 1970, on the other side of the world, the American Apollo space program was in crisis with one of their spacecraft – *Apollo 13*. The crew of the spacecraft was fighting desperately to get their crippled spacecraft back to Earth after a catastrophic explosion in its service module caused their Moon mission to be cancelled. The Russians, who had been monitoring the communications between the Manned Spacecraft Center (MSC) and *Apollo 13*, shut down their transmissions to leave the frequency clear of any interference, but continued to monitor the ongoing crisis. Despite the continuing 'Cold War' that existed between Russia and America at the time, the Soviet Premier Alexsei Kosygin sent a message to US President Richard Nixon and the US government saying: 'I want to inform you that the Soviet Government has given orders to all citizens and members of the armed forces to use all necessary means to render assistance in the rescue of the American (*Apollo 13*) astronauts.'

Russia also immediately offered their help unreservedly, both at sea and in the air, in the event of the spacecraft going off course after re-entry and coming down in their part of the world. They even went as far as diverting two of their cargo ships that were sailing in the Pacific Ocean at the time into the expected landing area, to be available should their assistance be required. Fortunately the spacecraft and the three astronauts returned safely in the planned area, and the help of the Russians, or indeed other countries who had offered to help, was not required. However the incident brought home to both countries the need for co-operation if such a problem arose again, and it was probably this that helped accelerate the Apollo/Soyuz Test Project (ASTP).

The Russians continued to carry out space duration tests and on 1 June 1970, *Soyuz 9* attached to an R-7 carrier rocket, with cosmonauts Major General Andriyan Grigor'yevich Nikolayev and civilian engineer Vitali Ivanovich Sevast'yanov aboard,

lifted off the launch pad at Baikonur and carried out the longest endurance record for a manned space flight to date: 17 days, 16 hours, and 59 minutes. Their mission was to observe weather patterns, geological objects, snow and ice fields, as well as carry out astronavigation exercises using a sextant. During the mission the two cosmonauts played a game of chess with the head of the cosmonauts, Nikolai Kaminon, and fellow cosmonaut Viktor Gorbatko. This was the first time any game had been played between participants in space and on Earth. The chess set aboard the Soyuz spacecraft had pegs and grooves to keep the pieces in place. Magnetic pieces had been considered but were rejected in case they interfered with the electronics. The match was declared a draw. *Soyuz 9* returned to Earth on 19 June landing on the Steppes of Kazakhstan.

The Russians were slowly accumulating a great deal of experience in time spent in space and the effects of weightlessness on the crews. The length of time the two *Soyuz 9* cosmonauts spent in space was significant, inasmuch as they suffered quite badly from the effects of living in a zero gravity in the cramped confines of the spacecraft. For over two weeks after their return they had difficulty in walking, stressing the point that cosmonauts/astronauts would have to carry out strict and rigorous muscle exercise whilst in space.

Two more missions to Venus were launched on modified SS-6 rockets in the 1970s, *Venera 7* on 17 August 1970 and *Venera 8* on 27 March 1972. Both missions were successful in sending back information on the atmosphere and the surface of the planet.

Then on 12 September 1970 the Russians launched *Luna 16*, another unmanned spacecraft attached to a Proton K/D carrier rocket. Its mission was to land on the

Lunakhod 1.

The Lunokhod that was successfully sent to the Moon. (Roscosmos)

Moon and bring back samples from the Sea of Fertility. After carrying out two orbits of the Moon, the spacecraft landed in the *Mare Fecunditatis* region and was the first unmanned spacecraft to land on the dark side of the Moon. After spending twenty-six hours and twenty-five minutes on the lunar surface, *Luna 16* lifted off and returned to Earth with her samples. The mission was a complete success.

On 8 October 1970 the FM-2 dummy was used again on a six-day orbital flight aboard *Kosmos 368*, a third generation photo-surveillance satellite, after which FM-2 was placed in the Polytechnic Museum in Moscow. One month later on 10 November 1970, *Luna 17* was launched attached to a Proton 8K82K with a Block D attachment rocket on another unmanned mission to the Moon. After a soft landing in the *Mare Imbrium*, the unmanned spacecraft released *Lunokhod 1*. The vehicle, resembling a large bathtub on wheels, was about the same size as a Volkswagen Beetle and was propelled on eight spoked wheels, powered by batteries that were charged by solar energy. It was equipped with several pieces of scientific apparatus, instruments, television cameras and a radio transmitter and receiver. Despite its ungainly appearance it was extremely successful, and during its operational life logged 6.5 miles (10.5 kilometres) and sent back 20,000 TV images and 206 high resolution photographs.

The launch of *Zond 8* on 20 October 1970 on a Proton K/D carrier rocket showed that the Russians had not given up on their efforts to reach the Moon. The mission had originally been planned as a manned flight, with cosmonauts Valery Bykovsky and

Nikolay Rukavishnikov on board, but confidence in the reliability of the spacecraft was still at an all-time low. Consequently it was decided to make it an unmanned, scientific, flyby mission. The launch, from an Earth parking-orbit platform, was successful and the flight to the Moon uneventful. The mission itself was a complete success, with numerous photographs being taken of the lunar surface and of Earth from the Moon. Once again problems with the attitude control system plagued the re-entry, and the spacecraft carried out a ballistic re-entry and splashed down once again in the backup area of the Indian Ocean. The spacecraft was recovered by the Russian recovery ship *Taman* close to the Chagos Islands and returned to Moscow. Although Vasili Mishin claimed that the ballistic re-entry was deliberate and under control, other sources say that the guidance system failed completely and re-entry was effected at over 20G, which could have been fatal had the spacecraft been manned. The planned launch of *Zond 9*, with cosmonauts Pavel Popovich and Vitali Sevastyanov on board, was cancelled. It was decided that too much time and money had been spent on this project, and in 1970, the L-1 program was cancelled.

The Russians then took the pace of the space race up a gear, when, on 19 April 1971, they launched *Salyut 1*, (DOS 1) the world's first space station, on a Proton-K rocket from Baikonur into a low orbit. This was to be the first of seven more stations, the final one being the module *Zvezda* (DOS 8), which became the core of the Russian segment of the International Space Station and remains in orbit. The DOS program arrived when Russia ran two programs simultaneously, a civilian scientific research one and a military one, *Almaz*, which was highly secretive.

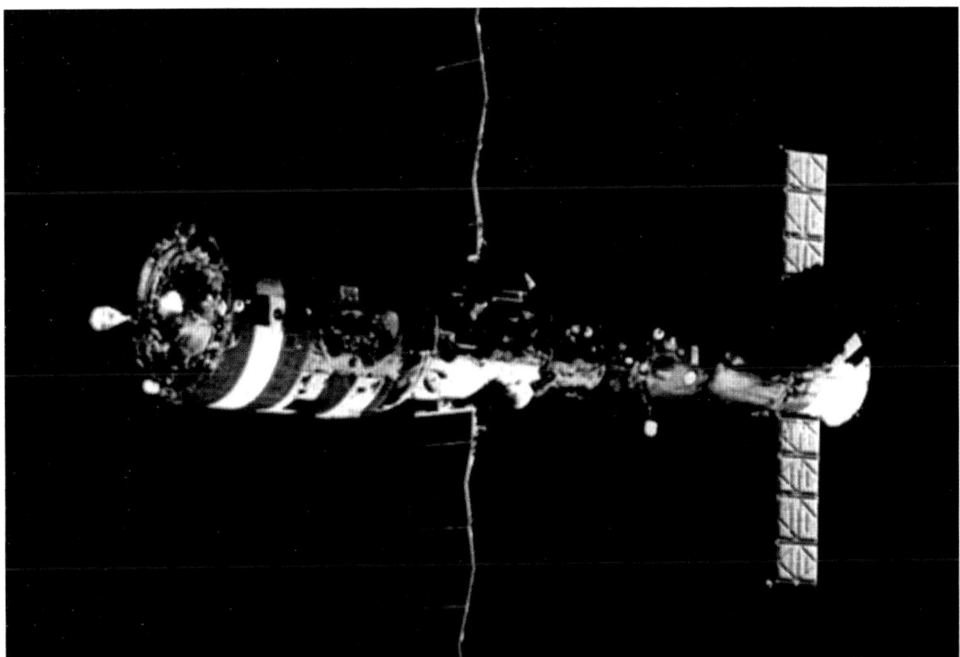

Salyut 1 in orbit. (Roscosmos)

Vladimir Chelomei. (Roscosmos)

Salyut 1 was 66ft long and 13ft in diameter, giving it a 3,500cu.ft. internal space. The station consisted of five components: a main section, a transfer compartment, two auxiliary compartments and the *Orion 1* space observatory. It was to spend nearly six months in orbit before it was replaced by *Salyut 2* and allow to burn up on re-entry. Known as DOS (Dologovremennaya Orbitalnaya Stanziya – Long Duration Orbital Station), the program was born out of the failure of the TsKBEM N1-L3 program and the fact that the only working spacecraft at the time was the Soyuz model. The only other operational program active at the time was the highly secret *Almaz* military space station project, run by the OKB-52 organization under the guidance of Vladimir Chelomei. There were also a number of military experiments carried on the space station, which included the highly classified Svinets radiometer, the OD-4 optical visual ranger and the Orion ultraviolet instrument for identifying rocket exhaust plumes.

The first cosmonauts to visit the *Salyut 1* space station were Vladimir Aleksandrovich Shatalov, Aleksey Stanislovovich Yeliseyev and Nikolay Nikolayevich Rukavishnikov. Their spacecraft, *Soyuz 10*, was the first of the upgraded 7K-OKS model and featured a new probe and drogue docking mechanism, giving it internal crew transfer capabilities when attached to a space station. It was launched on a Proton-K rocket on 23 April 1971 and attempted to carry out a hard docking with the space station over a period of five and a half hours. Officially no reason was given for the short duration of the mission, but it was discovered later that when Shatalov tried to engage with the space station, the docking mechanism on the spacecraft failed and only partially locked on. The problem was made worse when the failed docking mechanism wouldn't unlock, preventing the spacecraft from undocking from the station. During the five and half hour rendezvous, the air inside the *Soyuz 10* spacecraft became toxic and Rukavishnikov became unconscious. After a number of attempts the two spacecraft finally separated and the crew were able to return to Earth, landing near Karaganda, Kazakhstan. A post-flight investigation showed that the crew had no instrument that could provide them with the angle and range information necessary for a manual docking.

It wasn't until 6 June 1971, that the first cosmonauts went aboard the *Salyut* space station, when *Soyuz 11* was launched on a Soyuz rocket, with cosmonauts Georgy Dobrovolsky, Vladislav Nikolayevich Volkov and Viktor Ivanovich Patsayev aboard. The mission wasn't without incident: just two weeks into their stay on the space station, an electrical cable in one of the station sections caught fire and although it was quickly extinguished, it was with great difficulty that Chief Designer Vasili Mishin was able to

The crew of Soyuz 11 (Georgy Dobrovolsky, Vladislav Volkov, and Viktor Patsayev. (Roscosmos)

calm the three cosmonauts down and persuade them to continue their mission. It has to be remembered that the experience of living in a weightless environment in a relatively small, cramped 'metal container' was completely new to the three cosmonauts and any problems could become magnified and very concerning. There were a number of other problems that caused the mission to be terminated a week earlier than had been planned. After spending over three weeks in the orbiting space station and carrying out a number of medical experiments, investigations and astronomical observations using the *Orion* astrophysical observatory, the crew returned to their spacecraft and headed back to Earth. As their spacecraft entered the Earth's atmosphere everything was confirmed with the three cosmonauts that all was well, then they entered into the communications black-spot. The parachutes deployed automatically and the spacecraft touched down in the Karaganda region of Kazakhstan on 29 June. Within minutes a helicopter had landed alongside the spacecraft, but when the capsule was opened the crew were found to be in a collapsed state. Desperate attempts were made to resuscitate them, but they were all declared dead at the landing site.

It was discovered later that a ball joint in an exhaust valve had been accidentally dislodged, decompressing the capsule. Despite desperate attempts by Patsayev, who tried to plug the leak with his finger, the air in the capsule had leaked out. None of the cosmonauts were wearing pressure suits, which would have saved their lives. This was a dreadful blow to the Russian space program and it was to be more than two

Above: The ill-fated Soyuz 11 being rolled out onto the launch pad.

Left: Medical staff attempting to revive the *Soyuz 11* cosmonauts after they were discovered lifeless in their spacecraft after landing. (Roscosmos)

years before any cosmonauts were to enter space again. Chief Designer Vasili Mishin later stated that the crew could have used a manual drive to close the valve, but either they forgot or had never been shown this method. The incident caused a great deal of concern amongst the other cosmonauts, who were becoming increasingly worried about the number of problems that were constantly cropping up during missions. A new lightweight spacesuit, the Sokol, was developed and issued for all future spaceflights. In addition an environmental control unit was fitted and designed to pump air into the spacecraft in the case of a leak.

The launch of N-1 *L-6* rocket from Baikonur on the 26 June 1971 raised the hopes of another successful launch and a further step towards placing a cosmonaut on the Moon.

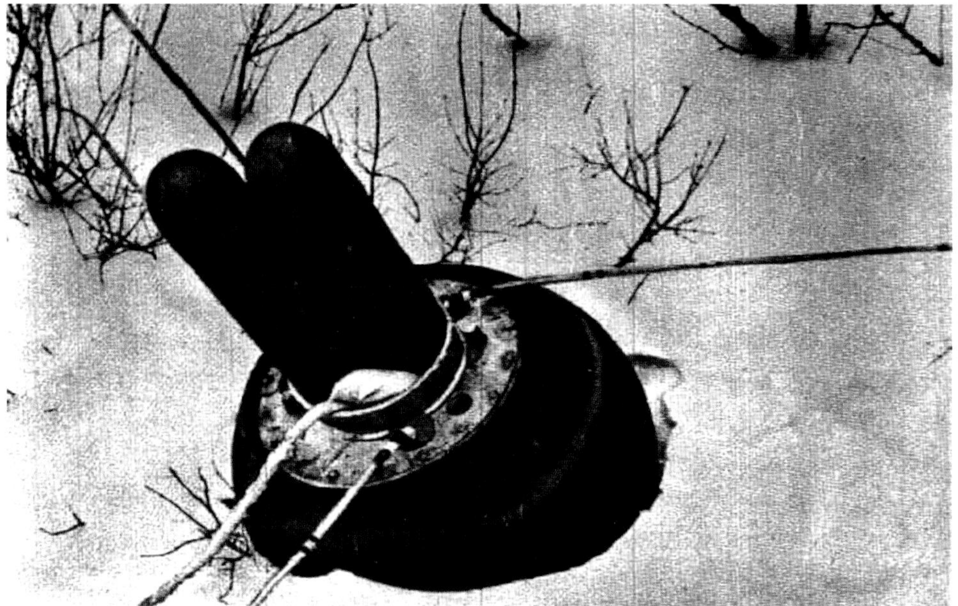

Luna 20 containing the first lunar samples, having just returned from the Moon.

Powered by thirty engines, the rocket lifted off the pad and for the first five seconds everything was normal, but then the rocket started a slow axial rotation. As the speed of the rocket increased so did the speed of the rotation and an automatic engine shutdown signal was transmitted. There was no response and by this time the rocket had rolled to sixty degrees, seconds later the engines shut down and the whole mission disappeared into a ball of flame as it hit the ground, some thirty kilometres from the launch pad.

The Russians launched another unmanned lunar probe, the Lavochkin built *Luna 18*, on 2 September 1971 on a Proton K/D rocket. After completing fifty-four lunar orbits, it attempted to land near the Sea of Fertility by firing its braking rockets, but on landing all communication with the probe ceased. It is believed that the probe crashed into the Moon's surface and was destroyed. It had been intended to collect samples of the lunar surface, but because of the spacecraft's demise this was abandoned. The Russians later said that the spacecraft had landed in an extremely rugged area, causing it to be damaged. Not to be discouraged, another probe was launched on 28 September, the Lavochkin-built *Luna 19* on a Proton K/D rocket. This mission was successful and it carried out geophysical research of the Moon's gravitational field and sent back even more photographs of the surface. Initially it had been intended to place the spacecraft into a low orbit to carry out the imaging of the lunar surface, but a faulty gyro placed it in an even higher orbit causing the high-resolution imaging and radar altimeter plans to be cancelled. The cameras were given a new role to take panoramic images of the lunar surface. After orbiting the Moon for over a year and completing 4,000 orbits, contact with the spacecraft was lost and it was assumed to have gone into a decaying orbit and crashed onto the lunar surface.

Then on 14 February 1972 *Luna 20* was launched on a Proton K/D rocket from Baikour. This was another one of the most ambitious projects carried out by the Russians to date, as the unmanned spacecraft was designed to land on the surface, retrieve some samples from beneath the lunar surface and return them to Earth. After carrying out a number of orbits of the Moon, the spacecraft landed in the mountainous area between the Sea of Fertility and Sea of Crisis on 21 February. An Earth-operated drill started to bore into the lunar surface to a depth of 35cm. Samples were obtained and transferred to a container that was hermetically sealed and readied to be returned to Earth. With the samples collected, the ascent stage of the spacecraft lifted off the Lunar surface on 22 February and returned to Earth. The tiny capsule touched down safely on an island in the Karkingir River just outside of the town of Jezkazgan in Kazakhstan. The result of the analysis of the samples showed that the area consisted of anorthosite, which contrasted greatly with samples obtained by *Luna 16* from the Sea of Fertility, which was primarily basaltic rock.

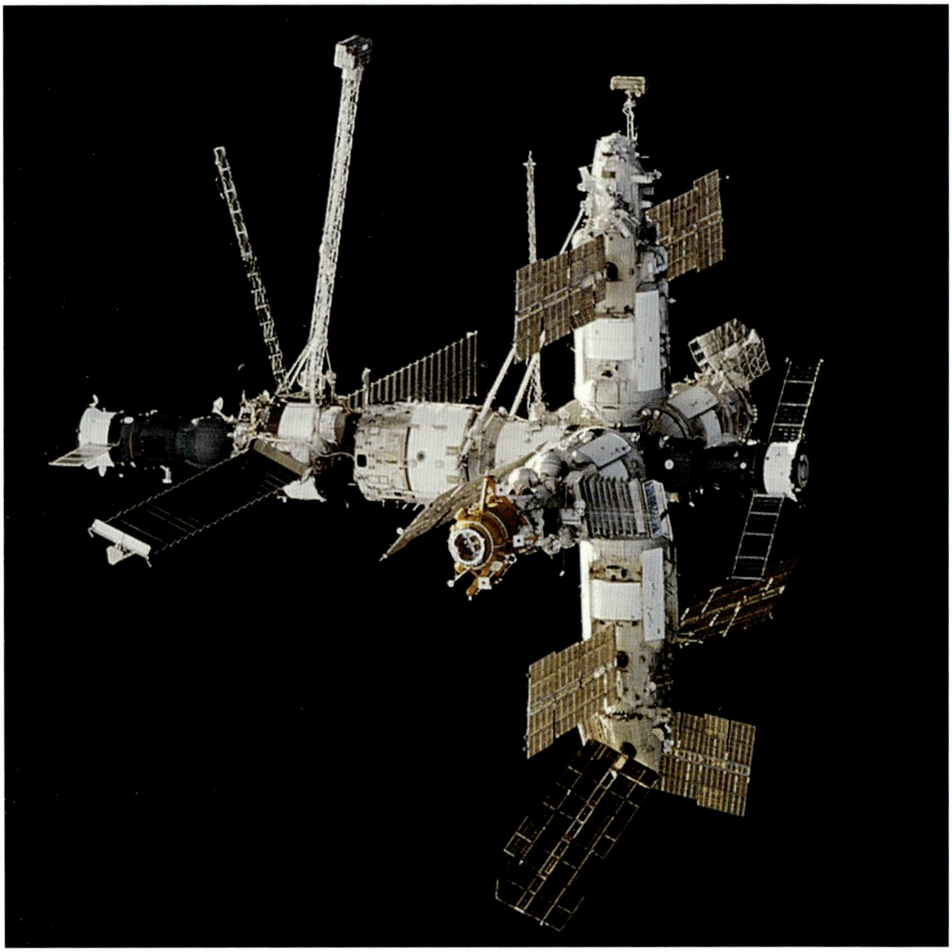

The Russian space station *Mir* showing all the 'add-ons'. (NASA)

Unmanned test flights continued with the launch of *Kosmos 496* on 26 June 1972. This was a test of a redesigned 'ferry' spacecraft that involved changes to the Salyut/Soyuz hatch, although no attempt was made to dock it with the space station. The spacecraft landed six days later in Kazakhstan on 2 July.

The development of the ill-fated N-1 rocket continued after the demise of N-1 *L-6*. In August 1972, N-1 *L-7* was on the launch pad undergoing the first of its tests, but the first of the dry runs revealed seventeen serious problems in the equipment systems and over 100 in the telemetry systems. The introduction of four roll-control engines solved the roll problem that had caused the N-1 *L-6* to be destroyed. Vasili Mishin, who was in overall control of the project, was ill at the time and so was unable to approve any launch date. The way things were progressing and the number of faults that were appearing would have made any launch date almost impossible to predict.

A second lunar rover, *Lunokhod 2*, was sent to the Moon on 8 January 1973. Launched on a Proton 8K82K rocket with a Block D attachment, the spacecraft *Luna 21* carried out forty orbits of the Moon before touching down in the Le Monnier crater between the Taurus Mountains and the Mare Serenitatis. Just three hours after touching down, *Lunokhod 2* rolled out on to the Lunar surface. This was an improved model of *Lunokhod 1* inasmuch as it had an additional TV camera mounted, an improved eight-wheel traction system, together with a large number of scientific instruments. Such were the improvements, that in the first day *Lunokhod 2* travelled farther than *Lunokhod 1* did in its entire mission. Disaster struck the mission however on 9 May, when the rover rolled over into a crater. Its solar panels and radiators were covered in thick lunar dust, which eventually caused the shutdown of power. Despite desperate attempts to recover the rover, it was decided to abandon the mission. Prior to this it had been relatively successful and over 80,000 TV images and 86 panoramic photographs had been sent back to Earth. It was revealed later, that one of the Russian engineers involved in the mission had been given photographs which had been taken by an American spacecraft that had orbited the Moon prior to *Apollo 17* landing in the area and had been used by the *Lunokhod* 'driver' to help navigate the rover.

The Orbital Piloted Station (OPS) program came to fruition in 1973 with the launch of *OPS-1 (Salyut-2)* space station. The program had been started in 1964 and was given an unlimited budget. So secret was the program and its location, that in the heavily guarded building where the space station was being assembled, even the air was controlled by a self-contained environmental control system, powered by its own independent diesel-electric generators. This was a far cry from the basic facility, with its frequent power failures, that was given to Vasili Mishin when he was involved with the development and construction of the N-1. The *OPS-1* space station was twenty metres in length and had a maximum diameter of four metres. The habitable area of the station was ninety cubic metres and consisted of four compartments. The first compartment was the transfer tunnel, which connected directly with the Soyuz spacecraft, the second and main compartment consisted of all the instruments and consoles, the third contained all the communication equipment, power supply, life support systems and auxiliary equipment. The engine, batteries, reserve supplies and water, were in the fourth compartment. Power from the batteries was acquired from the two double sets of solar panels that were situated externally at either end of the station. Between 1973 and 1976 there were three OPS missions: the first, *OPS-1*, was launched on a Proton rocket on 3 April 1973 and given the name *Salyut-2* so as to appear as a satellite and not to alert the West to the fact that it was a military space station. The Russians were still in experimental mode and the development of the *Salyut* space station and its obvious potential caused the Americans to think seriously about developing and constructing their own. Having said this, the launch of *Salyut 2*, (DOS 2) on 3 April 1973, was to be another setback in the Russian space program. The second of the Salyut space stations was in fact an *Almaz* military space station, launched under the cover of a scientific *Salyut 2*.

The space station was 47.7 ft long and 13.6 ft wide, giving it an internal volume of 3,200 cu.ft. The crew was due to launch ten days later, but three days after launching, the spent third stage suddenly exploded sending out a cloud of debris, some of which hit the space station, damaging the solar arrays. This caused the launch of the crew who were to man the space station to be delayed. Just prior to the crew being sent to the space station, a problem arose with the stability sensors and the space station began to gyrate violently to the point that it became unstable. Then, just ten days after it had been launched, ground control was made aware of a loss of pressure in the space station. At first it was said that one of the lines in the station's propulsion engines had burst and burnt a hole into the hull. Findings later proved that it had been debris from the spent Proton rocket which had exploded after separating, causing a piece of shrapnel to burst through the pressurized station's hull. Tracking stations from around the world started detecting anomalies coming from the area where the Salyut space station was supposed to be settling into orbit. It was quite obvious to those tracking the space station that it was in trouble and coming apart. The authorities denied that there was a problem and said that *Salyut 2* was settling into its pre-arranged orbit. Then on the 25 April tracking stations in North America announced that the *Salyut 2* space station was breaking up despite what the Russians were saying, and on 28 May what was left re-entered the Earth's atmosphere and was burnt up. Fourteen days later the tracking stations spotted another object being launched and then monitored it as it went into a very low orbit. It was suspected that it was an attempt to launch another space station, but this was flatly denied by Russia, who said that it was a 'test satellite', and it too burned up in the atmosphere eleven days later.

Another unmanned Soyuz spacecraft, *Kosmos 573*, was launched on 15 June 1973 to test the spacecraft without solar arrays. The results are not known, but it landed safely two days later in Kazakhstan. Meanwhile after a break of over two years, the manned Soyuz program re-started with the launch of *Soyuz 12* from Baikonur on a Soyuz rocket on 27 September 1973. The crew of Vasiliy Grigo'yevich Lazarev and Oleg Grigor'yevich Makarov were the first to venture into space since the tragic deaths of the crew of *Soyuz 11* two years previously. The primary mission of this spaceflight was to evaluate the newly designed Soyuz 7K-T spacecraft, and after two days in space they returned safely to Earth on 29 September, landing 440km south west of Karaganda, Kazakhstan.

Meanwhile another unmanned flight took place with the launch on 30 November 1973 of *Kosmos 613*. This was a long duration test of the Soyuz 'ferry' spacecraft in preparation for long stays whilst attached to a space station. *Kosmos 613* returned to Earth on 29 January, landing in Karaganda, Kazakhstan

The success of the *Soyuz 12* mission caused the next flight to be brought forward, and on 18 December 1973 just three months later, *Soyuz 13* blasted off the launch pad at Baikonur on a Soyuz rocket and into orbit. The crew, cosmonauts PetrIl'ich Klimuk and Valentin Vital'yevich Lebedev, carried on board their spacecraft the large Orion-2 astrophysical camera. They carried out astrophysical observations of stars in the ultraviolet range and spectrozonal photography of certain areas of the Earth. In addition to these experiments and investigations, they continued to test the systems aboard the new spacecraft. After nearly eight days in space the spacecraft returned to Earth and landed in the middle of a snowstorm 200km south-west of Karaganda, Kazakhstan on 26 December 1973. The rescue team arrived sometime later and were surprised to see the crew alive, so severe was the weather at the time.

On 3 April 1974 the unmanned Soyuz spacecraft *Kosmos 638* was launched to test an APAS-75 androgynous docking system in preparation for the forthcoming ASTP (Apollo/Soyuz Test Project). When the air was released from the orbital module just before re-entry, it was discovered that it caused a number of unexpected motions to the spacecraft. Although the spacecraft landed safely two days later, it was the cause of some concern, so it was decided that the second test, *Kosmos 672*, would be unmanned.

Space control ship *Cosmonaut Yuri Gagarin*.

In the meantime the second of two unmanned Lunar orbiters, *Luna 22* (the first was *Luna 19*), was launched on a Proton K/D rocket from Baikonur on 29 May 1974. The mission was to orbit the Moon and carry out a study of the Moon's magnetic field, surface gamma ray emissions, gravitational field, cosmic rays, micrometeorites, the composition of Lunar surface rocks as well as photographing and mapping the surface. For the next eighteen months *Luna 22* performed flawlessly, changing orbital heights throughout, sending all the details back to Earth and was only concluded in November 1975 when the manoeuvring propellant was exhausted. It finally went into a decaying orbit and crashed into the lunar surface. It was one of the most successful missions to date.

As the year progressed and problems were solved, only to be replaced with fresh ones, it became increasingly obvious that pressure from above was aiming to get the whole N-1 project cancelled and in August 1974 it was. Over 6 billion rubles had been spent over the previous 17 years to try and perfect the N-1 rocket but to no avail. Now work was to start to find a replacement.

A second Almaz military space station '*Salyut-3* (DOS 3)' was launched on 25 June 1974 and placed in orbit. On 3 July 1974 *Soyuz 14* was launched on a Soyuz rocket carrying the first crew of *Salyut 3*, Pavel Popovich and Yuri Artyukhin. The last 100 metres of their approach, and subsequent docking, was carried out manually, such was the reluctance to trust the automatic system. The improved spacecraft and the experience obtained were placing the Russians well ahead of the Americans. *Salyut 3* space station now had three sets of fixed solar panels grouped around the core instead of having two sets of solar panels that opened like wings fore and aft. The two cosmonauts installed four cameras, two in the lower part and two in the upper part of the station, and a solar telescope where the operator stood at its base and operated it by means of a control panel. One of the closely guarded secrets was the installation of a space cannon. The Almaz space station became the first, and only, space station to be fitted with a 'defensive' weapon when they installed a Nedelman-Rikhter R-23mm cannon originally installed in the Tupolev Tu-22 bomber. The defensive weapon, as it was called, was a 37-pound, rapid-fire cannon, that was capable of hitting targets over two miles away. The cannon was capable of firing 200 gram shells with a velocity of 1,500 miles per hour at a rate of between 1,800-2,000 shots per minute. To aim and fire the gun, the whole station would have to be turned to face the target, which in reality severely restricted its use. The remainder of their mission was a highly classified military one with some state-of-the-art spy equipment (cameras, radar, etc). No other details were made available. The space station was scheduled to fly seventeen orbits of the Earth per day, however seven of the orbits did not pass over Russia, causing a major problem in communications. To resolve this, two tracking ships, the *Cosmonaut Yuri Gagarin*, which was stationed in the Atlantic off Sable Island, Canada, and the *Cosmonaut Vladimir Komarov*, which was stationed off Cuba, were, between them, able to give the space station the additional coverage required to keep constant communication throughout the mission. Just before returning to Earth, the crew, it is said, tested the on-board cannon by destroying a test satellite that had been launched earlier. There is no evidence to support this, but if

Space control ship *Cosmonaut Vladimir Komarov.*

true, this would have been the first time that any weapon had been fired in space. It is an interesting thought that if one of the cannon shells had missed the target, it would possibly continue to travel through space for hundreds of years and millions of miles before friction caused it to stop.

After spending sixteen days checking the space station out and making sure that it was ready for the next crew, the two cosmonauts undocked and returned to Earth on 19 July 1974, landing within two kilometres of the estimated landing area that was 140 km southeast of Dzhezkazgan, Kazakhstan.

The second test of the ASTP docking system took place on 12 August with the launch of the unmanned *Kosmos 672* which was a complete success and was safely recovered on 18 August when the spacecraft touched down in Kazakhstan.

One month later on 26 August 1974, *Soyuz 15* lifted off the launch pad at Baikonur on a Soyuz rocket with cosmonauts Lev Stepanovich Demin and Gennadi Vasil'yevich Sarafanov aboard. Their primary mission was to rendezvous with the *Salyut-3* 'scientific' station and carry out a series of investigations and experiments. The mission was a total failure, as they were unable to dock because of a malfunction in the electronics in the Igla docking system. They also did not have sufficient fuel to carry out repeated attempts at manually docking with the station. This raised serious concerns about the latest Soyuz 7K-T spacecraft mainly because of its lack of reserve propellant and electrical power. It was also realised that the Igla docking system was in dire need of an overhaul and this was not going to happen anytime soon as the space station was coming close to the end of its operational life. The *Soyuz 15* spacecraft returned to Earth on 28 August, landing close to Tselinograd, Kazakhstan.

The space race, although on the face of it appearing to create a chasm between the United States and Russia, did in effect draw the two countries closer in some ways. Before the *Apollo 13* incident, President Kennedy had written to the Russian Premier at the time, Nikita

The crew of *Soyuz 16*, L-R: Anatoli Filipchenko and Nikolai Rukavishnikov (Roscosmos)

Khrushchev, about the setting up of a world weather satellite system and the exchange of acquired data concerning the Earth and the space surrounding it. This brought some limited response from the Russians, but after the *Apollo 13* incident, it was decided that in the event of any spacecraft, irrespective of nationality, getting into difficulties, there should be some common agreement that would enable either country to help set up rescue missions.

Discussions regarding a joint American/Russian space mission moved a step further when on 2 December 1974, *Soyuz 16*, with cosmonauts Anatoliy Valis'yevich Filipchenko and Nikolay Nikolayevich Rukavishnikov aboard, lifted off from Baikonur on a Soyuz-U rocket and into space. The crew carried out tests on the modified onboard systems that had been modernized to meet the requirements for the forthcoming ASTP mission. They also tested the androgynous docking system by retracting and extending a simulated 20kg American docking ring. In addition they also tested the modified environmental systems, an improved control system, the radar docking system and new solar panels. With almost all the tests complete, just prior to returning to Earth the crew jettisoned the docking ring using explosive bolts to test the emergency measures in the event that the capturing latches ceased to work during the ASTP mission. *Soyuz 16* returned to Earth on 8 December after a successful mission and landed outside Arkalyk, Kazakhstan.

On 26 December 1974 *Salyut 4* (DOS 4) space station was launched from Baikonur on a Proton three-stage rocket and placed in orbit with an apogee of 355 kilometres and a perigee of 343 kilometres. Sixteen days later on 10 January, *Soyuz 17* on a Soyuz rocket was launched with cosmonaut Alexi Gubarev and Georgi Grechko aboard and docked with the space station. As the two cosmonauts entered the space station, a sign that read 'Wipe Your Feet' greeted them, left by the space station's ground preparation team. The two cosmonauts carried out the usual checking and testing of all the on-board systems, including the testing of a piece of equipment that allowed the cosmonauts to obtain drinking water from the cabin atmosphere by condensing it. The work rate of the cosmonauts varied between fifteen and twenty hours a day and included a two and a half hour exercise program. Among the experiments carried out by the crew was the testing of communication equipment designed for tracking ships using a Molniya satellite. One piece of equipment that was introduced and operated very successfully was a teleprinter, which enabled the crew to work without being constantly interrupted to answer questions. A number of other experiments were carried out over the next twenty-nine days, then on 9 February the spacecraft undocked and returned to Earth, landing 110km northeast of Tselinograd, Kazakhstan. On 24 January 1975 the now defunct space station *Salyut-3* had re-entered the Earth's atmosphere over the Pacific Ocean and burnt up. One assumes that all the returning cosmonauts who had spent a number of days on the space station brought all their rubbish and unwanted materials back with them.

As if to tell the Americans that they were ahead again in the space race the Russians launched *Soyuz 18-1* on a Soyuz rocket on 5 April 1975. The two cosmonauts, Vasiliy Grigo'yevich Lazarev and Oleg Grigor'yevich Makarov, ran into problems soon after launch. Their primary mission was a rendezvous with the *Salyut 4* space station, but soon after the first stage had separated the second stage of the rocket fired, but failed to separate. With the second stage still firing the crew demanded that the flight be aborted, but the ground controllers could see nothing wrong in their telemetry signals. After much arguing the second stage was separated by a ground control command at a height of 119 miles (192 kilometres). An emergency re-entry was initiated, sustaining 20.6+G as it passed through the atmosphere and finally crashing into the Altai Mountains. The spacecraft then tumbled wildly down the mountainside coming to rest on the edge of a precipice. The only thing that prevented the spacecraft from hurtling over the side of the precipice and to almost certain death for the cosmonauts, was the fact that the parachute on the space capsule snagged on a tree.

Initially the crew thought that they had landed in China, so destroyed all documents that related to their intended military mission, but they were discovered sitting beside the capsule by some local people who spoke Russian. Because of the rough terrain and the weather conditions, the crew were not rescued until late the following day. Vasiliy Lazarev suffered internal injuries from the 20G re-entry and the tumble down the mountainside and never fully recovered from the experience – he was to never to fly again. Oleg Makarov also suffered some injuries, but his were relatively minor. All cosmonauts, after they had flown in space, were entitled to a spaceflight bonus of 3,000 rubles, but because they had not completed their mission, Lazarev and Makarov were refused the payment. It took appeals right up to President Leonid Brezhnev to get the decision reversed.

A proposed joint mission between the two superpowers was put forward, with the link-up in space of an Apollo and a Soyuz spacecraft. After a great number of difficulties, language and trust being just two, it was realized that both countries were going to have to be completely open about their space technology and the designs of their spacecraft. With the failure of the *Soyuz 18-1* mission still the cause for some concern, *Soyuz 18* was launched on 24 May 1975 with the crew, Colonel Petr Il'ich Klimuk and Vitaliy Ivanovich Sevast'yanov, on board. After a two day flight they docked with *Salyut-4* and carried out a number of experiments and investigations during their sixty-three-day mission. As with all the Russian missions, information regarding the types of medical and biological experiments and investigations were extremely sketchy and most of the time non-existent. Although it is known they did carry out in-depth investigations of the constellations Virgo, Scorpio and Cygnus using the X-ray telescope, these were essentially military missions. The spacecraft returned to Earth on 26 July 1975, landing east of Arkalyk, Kazakhstan.

The results of the *Soyuz 16* ASTP mission were the final pieces of the space jigsaw that culminated on 15 July 1975 when *Soyuz 19*, on a Soyuz 2-1A rocket, lifted off the launch pad at Baikonur and into space. On the other side of the world on 15 July 1975 a Saturn 1B rocket, with an Apollo spacecraft on top, lifted off the launch pad at the Kennedy Space Center and into orbit. The Apollo/Soyuz Test Project (ASTP) crews consisted of Soyuz cosmonauts, Aleksey Arkhipovich Leonov and Valeriy Nikolayevich Kubasov, and Apollo astronauts, Tom Stafford, Vance DeVoe Brand and Donald 'Deke' Slayton. During the early stages on the program both sides were known to have had a number of differences regarding the spacecraft and criticized each other's engineering. It was accepted that Russian spacecraft were designed to be more or less automatic, citing the number of uncrewed probes they had launched, and that the manual controls in the Soyuz spacecraft had been designed to minimize the risk of human error. The Russians considered the Apollo spacecraft to be highly complex and dangerous. The Apollo spacecraft on the other hand had been designed to be operated by highly trained astronauts and had very limited automation. The director of the Manned Spacecraft Center (MSC), Christopher C. Kraft, voiced his opinion by saying: 'We in NASA rely on redundant components – if an instrument fails during flight, our crews switch to another in an attempt to continue the mission. Each Soyuz component however is designed for a specific function; if one fails, the cosmonauts land as soon as possible. The Apollo vehicle also relied on astronaut piloting to a much greater extent than did the Soyuz machine.'

Despite their early differences the Russian and American scientists and engineers managed to put aside their opinions and work together. The astronauts and cosmonauts themselves had a number of meetings, all of which it is said, were relaxed and extremely harmonious. Such was the interest in this historic meeting

that the launch of the Soyuz spacecraft was broadcast live around the world, something that had never happened before in the history of the Russian space program. In fact very few Westerners had ever witnessed the launch of a Russian spacecraft or satellite.

The two spacecraft, now in orbit, edged towards each other, then at a height of 140 miles (225 kilometres) above the Atlantic Ocean they met. Minutes later the two spacecraft docked together with a gentle bump and history was made. On the television screens in Mission Control, Houston, Texas and in Soyuz control at Kaliningrad, Russia, appeared the close-up of a hatch. Suddenly the hatch opened and the smiling face of Russian cosmonaut Alexi Arkhipovich Leonov appeared, then a hand stretched out and the face of American astronaut Tom Stafford appeared on the screen. The two men shook hands warmly and greeted each other. The Apollo/ Soyuz link-up cemented the friendship between the Russian and American astronaut/ cosmonaut fraternity that had sprung up during training. Later both crews got together and carried out the first ever in-space joint press conference.

The mission lasted nearly six days during which a number of experiments were carried out, including one experiment carried out by the Russians, that consisted of three small cylinders that contained metals that could not be uniformly mixed on the ground because of the Earth's gravity. On Earth the heavier of the metals, in their molten state, settled to the bottom before the mixture cooled and solidified. The cosmonauts took the experiment, loaded the cylinders into an electric furnace in the docking module and proceeded with the sequence of heating, melting and cooling the samples in the weightless conditions. The test samples were returned to the scientists to evaluate the results. There were a number of medical experiments also carried out covering mainly biological investigations.

When the two spacecraft rendezvoused in space they opened up another chapter in man's desire to explore the universe. The whole mission was a complete success for both technologies and for the world. The American astronauts said afterwards, that the construction of the Russian spacecraft looked more like a plumber's nightmare than a sophisticated spacecraft. But what was apparent was that there was a true *esprit de corps* between the astronauts and the cosmonauts and that East and West could work together in the furtherance of peace and harmony.

There was almost a serious problem for the Apollo crew, when, just before splashdown, the crew was gassed by nitrogen tetroxide that had escaped from the RCS (Reaction Control System) thrusters. Although Vance Brand was rendered unconscious and the remaining two astronauts very woozy, they quickly recovered and suffered no ill effects from the experience.

Alexi Leonov and Tom Stafford became life-long friends after the mission. Alexi Leonov became godfather to Tom Stafford's younger children, and on Alexi Leonov's death, it was Tom Stafford who gave the eulogy at his funeral in Moscow.

In Russia the Mikoyan-Guryevich Company (MiG) was still working on the development of an aerospace system, known as the SPIRAL OS. It consisted of a re-usable hypersonic launch aircraft, an expendable two-stage rocket booster and an orbital spaceplane. The hypersonic launch aircraft was in the format of a large arrow-shaped flying wing, powered by four turbo-ramjet engines. On top of this was placed the orbital spaceplane and the expendable two-stage rocket. The theory was that the launch aircraft would take the spaceplane and the two-stage rocket to a height of thirty kilometres. The launch aircraft would then release them and return to its base. The two-stage rocket would ignite and push the orbital spaceplane into space. The two-stage rocket would then be released from the spaceplane and left to burn up as it re-entered the atmosphere.

After a number of wind tunnel tests and design alterations, it was decided to end the project because of the massive financial drain it was creating. The fact that it was under-funded from day one did not help, but the SPIRAL OS cosmonaut-training program was ended and the spacecraft used to test analogue systems.

On 8 June 1975 another mission to the planet Venus was launched, when a Proton K/D rocket carrying the *Venera 9* lander, lifted off the pad at Baikonur. Four months later the lander descended onto the surface of Venus whilst the Sun was at its zenith and managed to send back a wealth of information as well as a number of photographs before ceasing to operate fifty-three minutes later. The photographs were the first to be sent from another planet.

Although the manned space flights were still the main focus, investigating the Moon had not been forgotten, and on 28 October 1975 *Luna 23* on a Proton K/D rocket lifted off the launch pad at Baikonur. This mission was to carry out a soft landing on the lunar surface, then drill down and recover samples to be returned to Earth. When controllers back on Earth tried to activate the drill, they found that it was not working and had been damaged due to a hard landing. With that part of the mission now out of the question, they decided to incorporate a new limited scientific program, which they did until the power supply ran out. In March 2012, photographs, taken by the NASA Lunar Reconnaissance Orbiter (LRO), showed *Luna 23* lying on its side, which would explain why the mission failed.

Another unmanned Soyuz spacecraft, *Soyuz 20* on a Soyuz-U rocket, was launched from Baikonur on 17 November 1975 with a number of biological experiments aboard. It also carried out a series of tests over the next three months in which the spacecraft carried out a number of dockings and undockings with the *Salyut 4* space station in preparation for the use of the unmanned Progress cargo spacecraft.

The Russian designer Vladimir Chelomei resumed work on his spaceplane, although in 1972 a design developed by NPO Energia called the Buran, which was very similar to that of the American Space Shuttle's Orbiter, had been accepted as the prime spacecraft in the development of a re-usable space shuttle system.

Whilst the arguments over the Buran had been going on, the Russians had been continuing with their Salyut program with the launch of the *Salyut 5* (DOS 5) space station on 22 June 1976. The first of the Soyuz spacecraft to visit the station and dock was *Soyuz 21*, with cosmonauts Colonel Boris Valentinovich Volynov and Lieutenant-Colonel Vitaliy Mikhaylovich Zholobov on board. The launch, on a Soyuz rocket from

The *Venera 9* lander that was sent to the planet Venus. (RKS Energia)

Baikonur on 6 July 1976, went ahead with no difficulty and entered orbit at a height of 170 miles (274 kilometres). On the second day in orbit the two spacecraft closed and docked. The mission, which was mainly a military one, was scheduled to last two months and coincided with a large military exercise being held in Siberia. A number of experiments, including the use of a hand-held spectrometer to study aerosol and industrial pollution in the Earth's atmosphere, had been carried out on board the orbital research station, when Zholobov became unwell. The illness had started after they had been in space for about thirty days and had got progressively worse. After forty-nine days in space, and the majority of the experiments and investigations completed, it was decided to return the two cosmonauts to Earth. Another problem reared its head when the crew tried to undock from the space station. As the small manoeuvring jets were fired to move the spacecraft away, the docking mechanism, which should have released automatically, jammed. After another orbit of the Earth, the crew, using emergency procedures, tried again; this time they only managed to loosen the latches. After a number of frantic messages to and from the mission control, a different emergency procedure was used which finally succeeded in releasing the spacecraft. *Soyuz 21* landed 124 miles (200 kilometres) south-west of Kokchetav, Kazakhstan on 25 August 1976. It was never officially disclosed what illness Zhobolov had suffered from, but a psychological breakdown was one theory and space sickness another, but what was indicative was that he never flew again. (Space sickness usually lasts from one to three days from the start of the mission.)

The third and last of the Lunar missions, *Luna 24*, was launched from Baikonur on a Proton K/D rocket on 9 August 1976. This was an attempt to extract a sample of lunar material from the previously unexplored *Mare Crisium* area. The spacecraft touched down safely on 18 August after orbiting the Moon for a number of days. Controlled from Earth, the lander deployed and extended its drilling arm and commenced drilling to a depth of two metres. It then extracted the sample and stowed it into a small capsule before sealing it. Then, on 19 August, the ascent stage of the spacecraft lifted off the lunar surface and headed back to Earth with the sample. Three days later the capsule parachuted down and landed south-east of Surgut, Siberia where it was recovered. This was the last spacecraft to make a soft landing on the Moon until China's *Chang'e 3* on 14 December 2013. It was claimed by Russian scientists in February 1978, that they had detected water in the sample that had been returned on *Luna 24*. Then in September the Russians decided to make use of one of their spare Soyuz ASTP spacecraft, and in place of the Apollo/Soyuz docking unit, they fitted a Zeiss camera. With civilian engineer Vladimir Viktorovich Aksyonov and Colonel Valeri Fedorovich Bykovsky on board the spacecraft, *Soyuz 22* lifted off the launch pad at Baikonur on 15 September 1976 on a Soyuz-U rocket. The mission of *Soyuz 22* was to test and perfect a number of scientific-technical instruments that could be used for studying the geological characteristics of the Earth's surface from space. When perfected, it was also argued that similar methods could be used when exploring other worlds for economic purposes. With all the experiments and scientific investigations completed, the spacecraft returned to Earth on 23 September 1976 and landed 93 miles (150 kilometres) north-west of Tselinograd, Kazakhstan.

In the meantime Russia was really progressing with its space station program and this was highlighted by the fact that while *Soyuz 22* was in space, *Soyuz 23* was already on the launch pad being prepared for another mission to *Salyut 5* space station. The two cosmonauts, Lieutenant-Colonel Valeri Il'ich Rozhdestvenskiy and Lieutenant-Colonel Vyacheslav Dimitriyevich Zudov, clambered into the cramped capsule that was to take them to their home for the next couple of months. On 14 October 1976 the *Soyuz 23* spacecraft on a Soyuz-U rocket lifted off the launch pad and into orbit to rendezvous with *Salyut-5*. As the spacecraft attempted to dock with *Salyut 5*, a fault was discovered with the main antenna of the Igla rendezvous system. After repeated unsuccessful attempts to dock the two spacecraft, the mission was aborted and the Soyuz spacecraft returned to Earth. But the drama didn't end there. As the spacecraft landed in Lake Tengiz it came down in the middle of a horrendous blizzard. Before the crew could jettison the parachute, it had dragged the capsule under the water. Frogmen managed to reach the sunken spacecraft and attach a flotation collar to it, but the rescue boat was unable to reach the crew because of the partially frozen surface of the lake. Divers then managed to get another line on the spacecraft and connect it to one of the helicopters, which then pulled it into the shore. After spending almost half the night trapped in their capsule, the two cosmonauts were released. When the rescuers reached them they were surprised to find the two of them still alive.

The next Soyuz mission was on 7 February 1977 when *Soyuz 24* was scheduled to dock with the *Salyut 5* space station. On 7 February, *Soyuz 24* lifted off the launch pad on a Soyuz-U rocket and into orbit around the Earth. Their mission, primarily a military one, was to carry out a series of scientific experiments very similar to those of *Soyuz 21*. This was one of the most successful Soyuz missions in terms of experiments and investigations accomplished to date. The flight was scheduled for seventeen days and within that time, the two cosmonauts, Viktor Gorbatko and Yuri Glazkov, accomplished as much as the crew of *Soyuz-21* did in fifty days. On 25 February 1977 the spacecraft landed just 22.9 miles (37 kilometres) north-east of Arkalyk, Kazakhstan.

Whilst the Americans continued with the development of their Space Shuttle, the Russians kept up their investigations in the Salyut scientific research complex. On 29 September 1977 the *Salyut 6* (DOS 6) space station was placed into orbit around the Earth by a Proton 8K82K rocket. It was made up of three cylinders connected by conical adapters and consisted of a transfer module, a work compartment, an intermediate chamber, equipment bay and a science bay. *Salyut 6* differed from the previous Salyut space stations inasmuch as it had two docking ports, one at each end of the station. Power was obtained from three solar panels, two of which were mounted laterally and the other mounted in a dorsal position.

The next flight to the Salyut space station was meant to be one of the milestones, inasmuch as it was to be the first to use the two docking ports on the station. The second docking port had been put in place to enable an unmanned cargo spacecraft called Progress to dock with essential supplies for the crew of the space station. This also meant that when a crew went aboard to carry out experiments and investigations they could be joined by another crew some months later who would then take over, thereby releasing the first crew so that they could return to Earth. Another crew would replace the remaining crew aboard the station, some months later, each crew overlapping the other. The living space had an area of 100 cubic metres, whilst the transfer module was bounded by a sealed cylindrical shell just two metres in diameter and a sealed cone-shaped shell. The conical part was fitted over the top of the passive cone-shaped docking mechanism (the active part of the mechanism was on the transport spacecraft). A hatch was fitted into the conical structure which enabled the crew to egress and ingress the station for EVA purposes whilst it was in space.

A large amount of equipment was fitted on the outside of the space station, along with handrails and anchoring points to which cosmonauts in their EVA spacesuits could attach themselves to the station whilst they carried out external servicing and repairs. There were also panels for studying micro-meteoric particles and the soiling of external optical surfaces. External television cameras, spherical tanks of the gas composition support system that contained a supply of compressed air, and sun and ion orientation sensors for the station attitude control system, were also mounted on the outside of the space station.

The transfer module, which housed the cosmonauts' spacesuits and attachments and control panels, was also the air-lock chamber. The module also had seven portholes, some of which were fitted with astro-orientation instruments that, together with their responding controls and control panels, maintained the orientation of the space station. The work compartment was made up of five posts. Post No.1 was the main control area for the primary control systems and was situated in the lower part of the compartment. There was just enough room for two cosmonauts to work side-by-side. The post also contained the re-generation cartridges of the gas composition support system and the cooling and drying units of the thermal control system. Post No.2, also called the astropost, was the astro-navigation and astro-orientation facility section. Between the two posts was the area where the cosmonauts had their meals, at a small table fitted with special appliances for heating food.

Post No.3 was equipped to operate all the scientific apparatus and was located in the large diameter section of the work compartment close to the far end of the post. As with all the posts, there were facilities for anchoring and communication. All the food supplies were kept in containers in the instrument section. At the far end of the upper section of the work compartment were the toilet and shower facilities, and close by were two air-lock chambers used for the removal of the biological waste products. They were collected in special containers and ejected into space where they burnt up in the atmosphere.

Post No.4 was located in the lower central part of the work compartment where the majority of the medical experiments were carried out, and also housed the photographic and television equipment. On both sides of No.4 Post were the cooling and drying units of the thermal control system, the electronic units of the station's Attitude and Motion Control System (AMCS) and the radio equipment. The AMCS was one of the most vital systems on the space station because without it the station would be continuously rolling and turning in an uncontrolled manner. The water re-generation and scientific apparatus control panels were housed in Post No.5.

There was a gap of some nine months before the next Soyuz flight '*Soyuz 25*' lifted off on 9 October 1977 from Baikonur on a Soyuz-U rocket with cosmonauts Vladimir Kovalyonok and Valeri Ryumin aboard. Their mission was to dock with *Salyut 6* and carry out a series of experiments aboard the scientific research complex. The launch was copybook right up until the docking procedure was initiated. When the crew attempted to 'hard' dock, the spacecraft would not lock on and retracted itself automatically. The crew made further attempts to 'hard' dock but without success. A 'hard' docking is required for all the electrical connections to be firmly locked in place. After repeated attempts to dock with *Salyut 6* the mission was aborted. The *Soyuz 25* spacecraft returned to Earth on 11 October after only two days in space and landed just outside Tselinograd, Kazakhstan. It was decided that for future missions, one of the cosmonauts must have flown before into space, thus ensuring at least one of them would be experienced. Why this decision was made is a mystery, because the problem was that it was the mechanism that was faulty, not because of the inexperience of the cosmonauts.

Soyuz 26 was launched on 10 December 1977 on a Soyuz-U rocket from the space center at Baikonur with Yuri Romanenko and Georgi Grechko aboard. This was a risky mission because of the problems *Soyuz 25* had experienced trying to dock with the forward port of *Salyut 6*, and because of this the crew of *Soyuz 26* docked with the aft docking port. After successfully docking with the space station, the two cosmonauts entered, powered it up and started their experiments and investigations, the first cosmonauts to do so.

A number of modifications had been made to *Salyut 6* and one of these was fitting an inward opening EVA hatch on the side of the forward transfer compartment, enabling it to be used as an airlock.

The compartment also contained two of the new semi-rigid spacesuits. This new type of Orlan-MKS suit, as it was called, had an all-metal body piece with flexible soft material for the arms and legs and also had greater flexibility than the earlier models. The joints had sealed bearings and articulated joints that corresponded with the arm and leg joints of the wearer. In the event of an emergency, the suit could be donned in under five minutes. It had a number of advantages over the usual EVA suit inasmuch as the hermetic sealing provided greater security and fewer sizes were required to fit the varying sizes of the respective wearers. Later the two cosmonauts got into their EVA suits and exited the space station through the airlock. The EVA lasted 1 hour and 28 minutes during which time they discovered that there was no fault in the forward docking system of the space station, therefore the fault had to lay within *Soyuz 25*'s docking mechanism. During the spacewalk to inspect some suspected damage to the primary docking port, thought to have been caused by *Soyuz 25* when it hard docked, Romanenko, who was not tethered to the spacecraft, missed his hold and started to float away. Fortunately for him, Grechko, who was tethered, stretched out as far as he could and managed to catch hold of him before he got too far from the spacecraft. Romanenko was in fact still tethered to the space station via the electrical and communications umbilical cord, however this was never designed to be used as a tethering cable and would have quite easily broken. A problem later manifested itself when one of the cosmonauts' face-plates began to fog up whilst on an EVA. At first it was thought to have been a one-off incident, but it was later discovered that there was a serious problem with the seal around the face-plate. Like all early mishaps and incidents on Russian spacecraft and space stations, problems like this were never admitted until many years later. With all household duties completed, the crew then started to prepare the station for the arrival of two more cosmonauts. On 20 January 1978 the space station was the first to receive supplies of fuel, equipment, instruments, food, water and materials from an unmanned cargo spacecraft – *Progress 1* – which was launched from Baikonur on a Soyuz-U rocket. After being unloaded, it was refilled with unwanted materials and waste over the next few weeks before being undocked on the 5 February and placed in a decaying orbit where it burnt up as it re-entered the Earth's atmosphere two days later. The cost of building and launching an expendable Progress cargo spacecraft was around $50 million, a staggering amount of money when one consider the regularity in which they were sent to the various space stations over the years.

One week later, on 27 January 1978, the crew of *Soyuz 27*, Vladimir Dzhanibekov and Oleg Makarov, lifted off Baikonur on a Soyuz-U rocket and joined the crew of *Soyuz 26*. *Salyut 6*'s propulsion system experienced a serious malfunction during the second crew residency and was not usable again for the remainder of the station's lifespan. As a consequence, it was limited to firing its attitude-control thrusters, and visiting spacecraft had to perform any orbital adjustments that were required. After each crew residency ended, it was necessary for a Progress cargo spacecraft to boost the station into a higher orbit, so it wouldn't decay until the next residency began. The four crew members worked together to carry out a series of experiments before the *Soyuz 26* crew returned to Earth aboard the *Soyuz 27* spacecraft on 16 January 1978. During their stay aboard the space station, it is said that a fire broke out which emitted so much black smoke that it almost caused the space station to be abandoned. Fortunately it was brought under control and the crew remained aboard. Once again the Russians denied this ever happened and it was to be many years before the information regarding this incident was released by the authorities, and even then they said it had been greatly exaggerated.

On 2 March 1978 the Russians made space history when an international crew arrived at *Salyut 6* aboard *Soyuz 28*. The two cosmonauts, Aleksei Gubarev of Russia and Vladimir Remek of Czechoslovakia (now the Czech Republic), were launched from Baikonur on a Soyuz-U rocket and carried out extensive tests on natural resources and also a photographic survey of the central and southern parts of the USSR, as it was then. They also made a large number of medical and biological research investigations. One of their investigations, called Extinkstia, involved the observation of the changes in the brightness of the stars as they disappeared below the horizon. The information gathered from this investigation was required for a study of the dust layer that had been formed by micrometeorites at a height of between 80–100 kilometres above the Earth. After seven days of intensive work Gubarev and Remek returned to Earth on 10 March 1978, landing in Kazakhstan. The space station was vacated on 16 March 1978, when the crew of *Soyuz 27* returned to Earth after spending 96 days in space.

During the next few months, preparations were made for another mission to the now empty space station, but it wasn't until 17 June that the Russian space station *Salyut 6* was visited again, when *Soyuz 29* left Baikonur on a Soyuz-U rocket with cosmonauts Vladimir Kovalyonok and Alexander Ivanchenkov aboard and docked with the space station.

The first thing that was required upon arriving aboard the space station was to prepare it for habitation. This meant activating the air re-generators, the thermal regulation system and the water recycling apparatus. The whole process took a week, during which time the two cosmonauts adapted to the weightless conditions they were going to have to work under for the next five months. The arrival of the unmanned

Crew of *Soyuz 30*, Pytor Klimuk and Miroslaw Hermaszewski. (Roscosmos)

Progress 2 cargo spacecraft, launched from Baikonur on a Soyuz-U rocket on 7 July, brought fresh supplies of food, water, oxygen and materials to the crew. When unloaded, the crew filled the cargo spacecraft with all their rubbish and unwanted materials and sent it into a decaying orbit around the Earth to be burnt up on re-entry. The next part of their schedule was the maintenance on the airlock and the normal housekeeping duties. They also installed a smelting furnace that was to be used by the next visiting crew during a three-day experiment. This was amongst a number of other experiments they had brought with them. The two cosmonauts then settled down to conduct a series of scientific investigation and experiments. Two weeks later they were joined by two more cosmonauts, Pytor Klimuk and Major Miroslaw Hermaszewski of Poland aboard *Soyuz 30*. Launched aboard a Soyuz-U rocket on 27 June 1978, their

Pytor Klimuk and Miroslaw Hermaszewski having just landed in their Soyuz spacecraft.

eight-day mission was to carry out a joint Russian/Polish investigation into the process of obtaining semi-conductor materials under weightless conditions, together with medical and biological research into how spaceflight affected human organisms. They also photographed the Earth's surface and oceans. With these experiments completed the spacecraft returned to Earth on 4 July 1978, landing in a Rostov state farm field 186 miles (300 kilometres) west of Tselinograd, Kazakhstan.

Another unmanned cargo spacecraft, *Progress 3*, was launched from Baikonur on a Soyuz-U rocket, arrived on 8 August 1978, with essential supplies of food, water and personal items together with a number of new experiments. Then two weeks later the crew of *Soyuz 31*, Valeri Bykovski and Lieutenant-Colonel Sigmund Jähn of the German Democratic Republic (GDR), arrived to join the crew of the *Salyut 6* space station, Vladimir Kovalyonok and Alexander Ivanchenko, on the 27 August. On 3 September 1978 the crew of *Soyuz 29* returned to Earth in the *Soyuz 31* spacecraft, landing in Dzhezkazgan, Kazakhstan and to a well-earned rest after spending over seventy-nine days in space.

The *Soyuz 31* crew were to spend the next sixty-eight days in space, and carried out extensive investigations and experiments covering a variety of medical-biological and technological programs. Visual and photographic studies were made of different areas of the surface of the world's oceans using a special multispectral unit made by the Carl Zeiss Jena optical works in the German Democratic Republic (GDR).

During this period the unmanned cargo spacecraft *Progress 4*, launched from Baikonur on a Soyuz-U rocket on 4 October 1978, docked with the space station with fresh supplies of food and water. After being unloaded it was refilled with waste and unwanted materials during the next few weeks before it was undocked on the 24 October and placed in a decaying orbit to be burnt up as it re-entered the Earth's atmosphere. With all their work and investigation completed, Bykovski and Jähn returned to Earth in the *Soyuz 29* spacecraft on 2 November 1978, landing in the Dzhezkazgan, Kazakhstan region. The *Salyut 6* space station had spent nearly thirteen months in space and orbited the Earth over 6,300 times.

The space station remained unmanned until 26 February 1979 when *Soyuz 32*, launched on a Soyuz-U rocket with cosmonauts Vladimir Lyakhov and Valeri Ryumin aboard, arrived and docked. For the next 110 days the crew carried out a series of investigations and experiments, coupled with some necessary repair work

Valeri Bykovsky and Sigmund Jähn, the crew of *Soyuz-31*. (Roscosmos)

on the station itself. During this time, *Progress 5* was launched from Baikonur on a Soyuz-U rocket and docked on 12 March, bringing much needed supplies for the crew and replacement parts for the space station. Once again the empty cargo spacecraft was filled with unwanted material and a quantity of contaminated fuel from the damaged Salyut propulsion system. On 5 April the cargo spacecraft was sent into a decaying orbit to be burnt up on re-entry.

Soyuz 33, with Nikolai Rukavishnikov and Georgi Ivanov aboard, was the next spacecraft scheduled to visit the space station, but, after launching from Baikonur on 10 April 1979 aboard a Soyuz-U rocket, it failed to dock with *Salyut 6*. This was the third time this had happened and no explanation was given at the time, but it was revealed some years later, that the main engines on the spacecraft had had a 'burn-out'. The speed at which the Soyuz spacecraft was approaching the space station gave the crew a strong indication that something was wrong. Rukavishnikov, as commander, made the decision to fire up the reserve engine, but if that failed they would stay in orbit for another six days until their oxygen ran out. The crew were left with two choices, either use the engine, if it fired, and dock with the space station and hope that a rescue craft could be sent, or use the engine to head back to Earth. The crew chose the latter and fired up the engine which had to fire for more than 90 seconds or they would become a satellite; it did, but for 213 seconds. The spacecraft landed 180 kilometres off course near Dzhezkazgan, Kazakhstan. The cause of the engine not firing was never released.

Soyuz 33 crew, Nikolai Rukavishnikov and Georgi Ivanov. (NASA)

Progress 6 was launched from Baikonur on a Soyuz-U rocket on 13 May and arrived at the space station on 15 May with additional supplies of food and water and some desperately needed parts. After docking in the aft docking port and being unloaded, the cargo spacecraft was filled with rubbish and unwanted items and undocked on 8 June. It was then placed into a decaying orbit where it burnt up the following day on re-entering the Earth's atmosphere. With the experiments and investigations completed the *Soyuz 32* crew prepared to leave the space station. In the meantime the next visit to the *Salyut 6* space station was by *Soyuz 34* on 6 June and was an unmanned spacecraft sent to collect Vladimir Lyakhov and Valeri Ryumin from *Salyut 6*. This was the first time that an unmanned 'lifeboat' spacecraft had been used and was a complete success. Earlier that year visits by a Hungarian, a Cuban and a Bulgarian cosmonaut had been cancelled on the grounds of 'cautious safety', obviously something that the Russian cosmonauts were not privy to. After docking in the forward docking station, the resident crew started to fill *Soyuz 32* with the results of their experiments and instruments after which they undocked the unmanned spacecraft and returned it to Earth where it landed safely just outside Dzhezkazgan, Kazakhstan. Meanwhile the two cosmonauts, Vladimir Lyakhov and Valeri Ryumin, got into *Soyuz 34* and moved it to the forward port docking station in preparation for the arrival of *Progress 7* which had been launched on a Soyuz-U rocket from Baikonur on 28 June, arriving to dock in the aft docking station. It also carried the KR-10 radio telescope that was to be fitted into the space station.

On 19 August 1979 with their work completed, the two cosmonauts clambered into their *Soyuz 34* spacecraft, together with a number of experiments, including the KRT-10 radio telescope, undocked from *Salyut 6* and headed back to Earth, landing southeast of Dzhezkezgan, Kazakhstan. Because the two cosmonauts had been living in a weightless environment for the past six months, special arrangements had to be made to remove them from their capsule using a system of slides and chutes. They had also become physically weakened, as one of the cosmonauts, when presented with the traditional bouquet of flowers, commented that the flowers felt like a ton weight when he accepted them. Both cosmonauts had lost weight including muscle wastage, but within seven days had replaced all of it.

CHAPTER SIX

International Crews

The station remained empty until the arrival of a new type of spacecraft in 1980, the unmanned Soyuz-T model – *Soyuz T-1*. It docked successfully and spent a number of months attached to the space station. It was undocked on 23 March and put into a decaying orbit to be burnt up on re-entry into the Earth's atmosphere. In the meantime a manned spacecraft arrived, *Soyuz 35*, which was launched from Baikonur on a Soyuz-U rocket on 9 April 1980. The two cosmonauts, Leonid Popov and Valery Ryumin, arrived on the *Salyut 6* space station on 10 April, and during the next 184 days, carried out household and maintenance duties plus a number of undisclosed military experiments. They were to establish a new record by spending 185 days in space, and during this period hosted four visiting crews, including cosmonauts from Hungary, Cuba and Vietnam.

Progress 7 was followed on 27 March 1980 by *Progress 8* bringing more supplies and allowing the crew to get rid of all their unwanted materials and rubbish. One month later *Progress 9* docked in the aft docking port of *Salyut 6*. This Progress spacecraft had been launched from Baikonur aboard a Soyuz-U rocket and was in fact a water tanker.

The fifth international crew, Valeri Kubasov and Bertalan Farkas (Hungary), to go aboard the space station, arrived aboard *Soyuz 36* on 27 May 1980. Launched from Baikonur on 26 May, the two cosmonauts were to spend the next sixty-six days aboard the space station, helping to carry out the normal maintenance and housekeeping duties, together with a number of undisclosed military experiments. The crew also had to unload the *Progress 9* cargo spacecraft and carried out the first-ever transfer of water between a water tanker spacecraft and a space station.

The first of the manned Soyuz-T spacecraft, *Soyuz T-2*, was launched aboard a Soyuz U rocket from Baikonur on 5 June 1980 and docked with the orbiting space station *Salyut 6* the following day. After they had docked with *Salyut 6*, the crew of Vladimir Aksyonov and Lieutenant Colonel Yuri Malyshev carried out testing of the on-board systems. Although the mission was very short, just under four days, it served its purpose in testing the new model Soyuz-T. The spacecraft landed on 9 June 1980 at Dzehezkagan, Kazakhstan.

The cargo spacecraft *Progress 10* was launched from Baikonur aboard a Soyuz-U rocket, docking in the aft docking port on 29 June and, after being unloaded, was undocked on 17 July and sent into a decaying orbit carrying unwanted materials and rubbish.

Above: Crew of *Soyuz T-2*, Leonid Kizim and Oleg Makarov, in training. (Roscosmos)

Below: *Progress MS-10* unmanned cargo spacecraft approaching the ISS. (NASA)

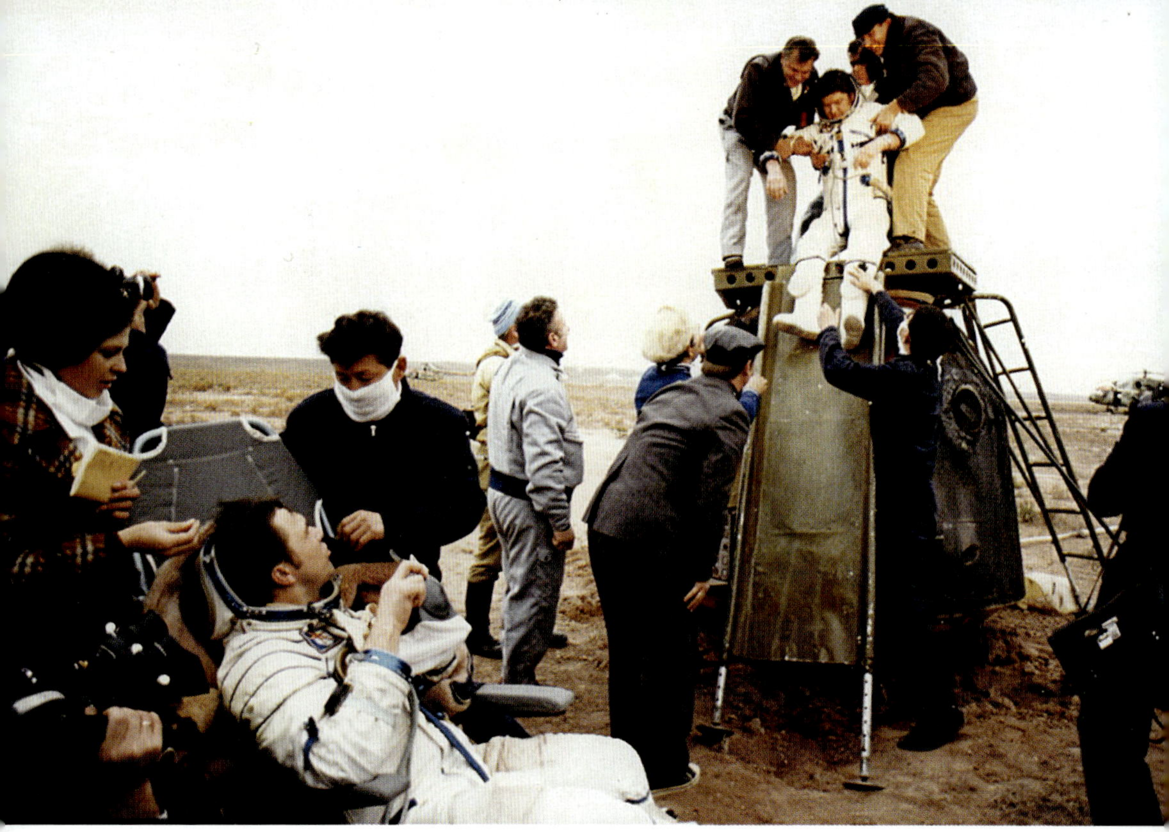

North Vietnamese cosmonaut, Pham Tuam, being helped from his spacecraft after landing. His fellow cosmonaut Viktor Gorbatko is seen relaxing in the foreground in his contoured couch. (RKS Energia)

One month later, on 23 July 1980, aboard a Soyuz U rocket, *Soyuz 37* was launched from Baikonur, with cosmonauts Colonel Viktor Gorbatko and North Vietnamese Air Force officer Lieutenant Colonel Pham Tuam aboard. After spending seven days aboard *Salyut 6* carrying out a series of experiments, both medical and engineering, they also took numerous photographs of Vietnam. The two cosmonauts returned to Earth in *Soyuz 36*, on 31 July, landing in the Dzhezkazgan, Kazakhstan region, leaving their *Soyuz 37* spacecraft as a return vehicle for Leonid Popov and Valeri Ryumin. Whilst in space, Colonel Tuam broadcast a number of political messages to the people of North Vietnam, extolling the virtues of the regime and thanking the Russian Communist Party for training him and sending him into space. However in North Vietnam, amongst the ordinary people, the messages were not so well received, causing a great deal of resentment which was widely voiced in the following manner: 'We have no rice, we have no noodles, so why are you in space Mr. Tuam?'

The crew of *Soyuz 36*, Valeri Kubasov and Bertalan Farkas, returned to Earth on 31 July using the *Soyuz 35* spacecraft, leaving their *Soyuz 36* spacecraft as a 'lifeboat' for the next crew who visited. They touched down near Dzhezkazgan three hours later. Valeri Kubasov was later part of the Apollo Soyuz Test Project (ASTP), when an American Apollo spacecraft and a Russian Soyuz spacecraft joined together in space.

On 28 September the two remaining cosmonauts, Valeri Ryumin and Leonid Popov, guided *Progress 11*, one of the unmanned cargo spacecraft that had been launched from

Baikonur on a Soyuz-U rocket, to the space station and into the aft docking port so that fresh supplies could be unloaded. The chemical batteries that powered the cargo spacecraft allowed it to operate in free flight for up to three days. With all the supplies stored away and all the rubbish and unwanted materials placed in the cargo spacecraft, *Progress 11* was then used to push the *Salyut 6* space station into a higher orbit, after which *Progress* undocked and placed in a decaying orbit around Earth to be burnt up on re-entry.

The movement of crews to the Salyut space station was becoming almost like a bus service. On 18 September 1980, *Soyuz 38* blasted off the launch pad at Baikonur aboard a Soyuz U rocket. On board was the seventh foreign cosmonaut to visit the space station *Salyut 6*, Arnaldo Tamayo Mendez of Cuba, the first Latin American in space, accompanied by Colonel Yuri Romanenko. During their seven-day stay they carried out a number of bio-medical tests concerning how the brain, blood circulation and eye functions of cosmonauts behaved under weightless conditions, and also researched into space sickness. The two cosmonauts returned to Earth on 26 September 1980 after completing their experiments, landing in the Dzhezkazgan area of Kazkhstan.

On 11 October the two cosmonauts, Leonid Popov and Valeri Ryumin, climbed aboard their spacecraft *Soyuz 37*, and headed back to Earth, landing close to

Unmanned Progress cargo spacecraft burning up as it enters the Earth's atmosphere. (NASA)

Dzhezkazgan, Kazakhstan. This had been Popov's and Ryumin's second visit to the space station and they had also set the longest duration record in space.

On 27 November 1980 *Soyuz T-3* lifted off the launch pad on a Soyuz-U rocket with cosmonauts Leonid Kizim, Oleg Makarov and Gennady Strekalov aboard. The following day the spacecraft docked with *Salyut 6* where the crew carried out repairs and household duties on the space station. This was the first Soviet three-man crew to be launched into space since the *Soyuz 11* disaster in 1971, when three cosmonauts lost their lives. During their time on *Salyut 6*, the crew carried out a number of electrical repairs and replaced a number of scientific items. They also carried out a series of biological experiments, scientific and technical investigations, before returning to Earth on 10 December after just twelve days in space, landing 81 miles (130 kilometres) west of Dzhezkazgan, Kazakhstan, leaving the space station empty.

It was to be three months before the *Salyut 6* space station was visited again. However, the cargo spacecraft *Progress 12* had been launched aboard a Soyuz-U rocket on 24 January 1981 and two days later docked remotely at the aft docking port, leaving the forward docking port for the next visiting crew. On 13 March, *Soyuz T-4* aboard a Soyuz-U rocket lifted off the launch pad with cosmonauts Vladimir Kovolyonok and Viktor Savinykh. After docking and firing up the space station, the two crew members spent the next couple of days getting the station habitable before unloading the supplies from the *Progress 12* spacecraft. Over the next couple of months the now empty cargo spacecraft was refilled with unwanted materials and rubbish. When this had been completed, the cargo spacecraft was undocked on 19 March and allowed to go into a decaying orbit before burning up on re-entry, leaving the docking port free for the arrival of the *Soyuz 39* crew. One of the experiments carried out by the crew was to replace the Soyuz spacecraft's probe with a Salyut drogue. This was to find out if a Soyuz-T spacecraft that was docked to the space station could be used as a rescue vehicle in the event that an approaching Soyuz-T spacecraft that was not fitted with a probe, could not dock with the space station and be unable to return to Earth. The results of the experiment are not known, as with almost all the experiments carried out aboard the space station. Ten days later, on 22 March 1981, *Soyuz 39* blasted off the launch pad on a Soyuz-U rocket with Russian cosmonaut Vladimir Dzhanibekov and Jugderdemidyin Gurragcha of the Mongolian People's Republic aboard. This was part of the INTERCOSMOS program designed to bring in and train cosmonauts from Russia's satellite countries. The two-man crew docked at the aft end of the *Salyut 6* space station on 23 March and, after carrying out a series of experiments and investigations, returned to Earth on 30 March 1981.

The last of the Salyut space stations, *Salyut 7* (DOS 7), was launched on 19 April 1982. It had originally been intended to be part of the forthcoming *Mir* program, but became the replacement for the rapidly becoming defunct *Salyut 6* because it had almost identical equipment and capabilities. The decision was made due to the continuing delays to the development of the *Mir* program, so it was launched as *Salyut 7*.

Soyuz 40 lifting off the launch pad at Baikonur. (RKS Energia)

Above: Buran spaceplane on top of the Antonov 225. (Roscosmos)

Below: Aerial shot of the Buran on top of the Antonov 225. (Roscosmos)

There were a number of problems affecting the old *Salyut 6* space station whilst it was in orbit, but the experience of the various visiting crews had enabled running repairs to be made.

Launched on 25 April 1981 on a Proton rocket, *Kosmos 1267 (TKS-2)*, which was almost the same size as the *Salyut 6* space station itself, carried out a number of orbital manoeuvres before finally docking on 19 June. It was said that the objective was to carry out experiments for the development of larger orbital space stations. *Salyut 6* continued to be the operational space station, and the last crew to visit before *Salyut 7* took over was *Soyuz 40*, carrying Leonid Popov and Dumitru Prunariu of Romania, who were launched from Baikonur on 14 May 1981. The two cosmonauts docked their spacecraft in the aft docking port of the *Salyut 6* space station the following day and, after conducting a series of investigations and experiments, undocked and returned to Earth one week later on 22 May 1981. The vast majority of the work carried out by all the visiting crews was the maintenance and housekeeping duties required to keep the space station operational. All experiments were carried out in addition to these necessary tasks, causing the cosmonauts to work very long hours. As with all the visits to the Salyut space stations, almost nothing is known of the experiments and observations that were carried out. On 26 May, the resident crew of *Soyuz-T-4*, Vladimir Kovolyonok and Viktor Savinykh, undocked from *Salyut-6* and returned to Earth, landing 78 miles (125 kilometres) from Dzhezkazgan, Kazakhstan. This was the last crew to inhabit the *Salyut 6* space station. It was during this period that in America a new type of space transportation was about to make its debut, the STS (Space Transportation System) Space Shuttle.

Soviet interest in a reusable spacecraft was revived with the development of the Buran (Snowstorm) spaceplane. The designers believed that their spaceplane would prove to be more reliable and more cost-effective than the present system. The final design of the Buran Space Shuttle Orbiter was almost identical to that of the American Space Shuttle Orbiter. Rumor had it that Soviet intelligence obtained the complete design specifications from someone in America. Initially Russian design engineers had dismissed the US design because of the large wings, and had carried out extensive aerodynamic tests whilst developing the Soyuz spacecraft. The results from these tests had shown that the large wings severely restricted the weight of the payloads that could be carried, because of the weight penalties incurred by the wings of the spacecraft itself. They favoured the lifting body design with virtually no wings. After a great deal of debate and considering a large number of different designs, it was decided, reluctantly, that the American design was by far superior to anything that they could come up with. As one NPO Energia designer/engineer said, 'There is no point in picking a different inferior design just because it is original, that just defeats the object.'

A new sub-contractor was chosen to build the spacecraft, the famous Russian aircraft manufacturer Mikoyan-Guryevich (MiG) who had been involved in the SPIRAL OS project. They created a new design bureau, Molinya, to create the Buran

Buran spaceplane on its launching vehicle being taken to the launch pad for its first and only flight. (RKS Energia)

spacecraft and carry out all the necessary tests. The Buran, also known as the MiG-105, was fitted with a ramjet engine, but required an assisted launch vehicle to take it into orbit. One of the first test pilots was cosmonaut Gherman Titov, the second man to orbit the Earth.

Spurred on by the progress the Americans were having in developing their Space Shuttle Orbiter and the interest being shown by the Russian Ministry of Defence, a series of five scaled unmanned models of the spacecraft, called the Bor (Bespilotniye Orbitalniye Raketoplan), were tested in secret. However some Australian fishermen saw a Russian Naval ship pulling a scale model from the sea after testing. The detailed description given by the fishermen was the first the Americans knew of a similar project to their own. With the development of the Buran, a number of free flights were made within the Earth's atmosphere, the Buran being fitted with four AL-31 turbofan engines and able to perform under its own power. It wasn't until 15 November 1988 that the Buran's only space flight took place when the unmanned spacecraft was attached to an Energia launch vehicle and blasted into space. After reaching space, it completed two orbits of the Earth before re-entering and landing using its fully

automated landing system. The spacecraft touched down safely at the Baikonur Cosmodrome 3 hours and 25 minutes later. On examination it was found that the Buran had lost just eight of its 38,000 thermal tiles.

Despite the hardships the people of Russia were suffering under Premier Brezhnev's regime, the Soviet Union pressed ahead with their space program. However there were a large number of dissenting voices about the amount of money being spent on the Buran program, but Brezhnev was more concerned with increasing military expenditure and prestige than the comfort of his own people.

Interest then turned to Chelomei's spaceplane LKS, which was very similar in design to the Buran but much smaller, so was selected as a backup, but was not completed until 1979. This included the construction of a full-scale mock-up which, according to some sources, was completed in less than a month and shown to the military hierarchy in an attempt to get the Buran project cancelled. The reverse happened. Later, in 1981, the development of the LKS was stopped and the whole project put on hold. In 1991 the workshops of NPO Mashinostroyeniye, where the spaceplane was still being developed, were broken into and everything, including the mock-up, plans and machinery, was destroyed. It is thought this was the result of the bitter arguments that Chelomei had had with the Kremlin hierarchy. Rumour was that it was the KGB who were behind the incident and there may have been some element of truth in that, but in Russia at the time, anything that dealt with state security was surrounded by absolute secrecy.

Despite the ongoing political pressures affecting other parts of the space industry, the space station program continued, with crews spending longer and longer in space. On 3 May 1982, *Soyuz T-5*, with cosmonauts Anatoli Berezovoi and Valentin Lebedev on board, blasted off the launch pad to rendezvous with *Salyut 7* and join up with cosmonauts Vladimir Kovalyonok and Viktor Savinykh of *Soyuz T-4*. After carrying out a series of medical and scientific experiments and the usual maintenance the *Soyuz T-4* crew returned to Earth, landing in Kazakhstan on 26 May after spending almost seventy-five days in space, leaving the *Soyuz T-5* crew to carry out the remainder of the experiments and repairs. On 25 June 1982, the crew of *Soyuz T-6*, Vladimir Dzhanibekov and Jean-Loup Chretien (France), joined the *Soyuz T-5* crew aboard *Salyut 7*. The Soviet/French crew arrived to carry out a series of scientific and medical experiments, but only stayed a matter of eight days before returning to Earth.

In the United States the R & D (Research and Development) Shuttle flights were nearing the end of their flight tests, the last to be launched being *STS-4* (*Columbia OV-102*).

The next to arrive at the *Salyut 7* space station was the *Progress 13 7K-13G* unmanned cargo spacecraft, which lifted off the launch pad at Baikonur on a Soyuz-U companion rocket on 23 May 1982 and delivered supplies and materials to the space station's resident crew. This was the first cargo spacecraft to visit the *Salyut 7* space station. Once again after unloading the cargo, the spacecraft was refilled over the next few weeks with rubbish and unwanted materials. It was undocked from

Above: Buran spaceplane touching down after its maiden flight. (Roscosmos)

Below: Buran spaceplane touching down after its one and only flight. (Roscosmos)

Buran spaceplane being 'de-gassed' after its flight. (Roscosmos)

the space station on 4 June and placed into a decaying orbit, and on 6 June it was burnt up as it re-entered the Earth's atmosphere. Another cargo spacecraft, *Progress 14*, was launched on a Soyuz-U rocket on 10 July with fresh supplies. The unmanned spacecraft docked with the space station on 12 July where it was unloaded. After all the waste materials and unwanted items had been placed aboard the cargo spacecraft, it was undocked on 10 August and placed in a decaying orbit where it burnt up on 13 August as it re-entered Earth's atmosphere.

On 19 August 1982, *Soyuz T-7* lifted off the launch pad and docked with the *Salyut 7* space station. The crew, Leonid Popov, Alexander Serebrov and Svetlana Savitskaya, were to spend the next ten days in the *Salyut 7* space station carrying out experiments and investigations. This was the first flight for Svetlana Savitskaya; she was only the second Russian female cosmonaut to go into space and the first to live on a space station. After docking, the crew carried out the usual test procedures before cracking the hatch and being greeted by the resident crew. They then settled down to their allotted tasks and the more mundane, but necessary, housekeeping duties.

On 29 July 1982, with *Salyut 7* now in orbit, the *Salyut 6* space station, together with *Kosmos 1267* still attached, was declared to be redundant. After spending five years in

Soyuz T-4 crew of Vladimir Kvalyonok and Viktor Savinykh inside their spacecraft. (Roscosmos)

space, *Salyut 6* was deorbited and allowed to gradually enter the Earth's atmosphere and burn up on re-entry over the Pacific Ocean.

The arrival of *Progress 15* on 20 September brought fresh supplies and equipment to the *Salyut 7* space station. Once again, after being unloaded, it was filled with discarded and unwanted materials and on 14 October it was undocked. Two days later it entered the Earth's atmosphere and burnt up on re-entry over the Pacific Ocean. Two weeks later, on 31 October, another cargo spacecraft arrived, *Progress 16*, with more supplies and experiments for the crew members on the space station. It was to remain docked until 14 December before being undocked and being placed in a decaying orbit.

After joining up with the space station and the crew of *Soyuz T-5*, the *Soyuz T-7* crew carried out a number of experiments before they returned to Earth on 23 November 1982. The *Soyuz T-5* crew returned to Earth on 10 December 1982, in the *Soyuz T-7* spacecraft, leaving the *Salyut 7* space station empty. On 7 March 1983, the empty space station was visited by an unmanned TKS ferry spacecraft called *Kosmos 1443 (TKS-3)* that had been launched aboard a Proton rocket from Baikonur and had been sent to collect experiment results from *Salyut 7* and hardware no longer required.

The problems of docking came back to haunt the Soyuz spacecraft on 21 April 1983. Launched from Baikonur on 20 April, *Soyuz T-8* on a Soyuz-U rocket, with

cosmonauts Alexander Serebrov, Gennadi Strekalov and Vladimir Titov on board, rendezvoused with the *Salyut 7* space station. The two spacecraft closed on each other but failed to dock, as the docking latches would not operate. After a number of attempts, it was decided to abandon the mission and the spacecraft returned to Earth on 22 April, landing near Arkalyk, Kazakhstan,

Cosmonaut Svetlana Savitskaya – the second female to go into space (Roscosmos)

On 27 June 1983, the Russians put another spacecraft into orbit on a Soyuz-U rocket, *Soyuz T-9*, with cosmonauts Vladimir Lyakov and Alexsander P. Aleksandrov aboard, to dock with the *Salyut 7* space station. With the docking problems *Soyuz T-8* had experienced fresh in their minds, the crew edged their spacecraft gently into the docking port and it was with great relief that the docking was successful and completed without incident. This was the fourth expedition to the *Salyut 7* space station and their first priority was to get it up and running, with the normal household duties. They then had to unload the 3.5 tons of cargo from *Kosmos 1443* which had arrived back in March. A number of medical and scientific experiments were carried out by the crew during their 149-day stay aboard the space station.

On 27 July the crew experienced a bang to the outside of the space station when an object struck one of the view-ports, creating a 4mm crater in the outer window, but fortunately it did not penetrate. They thought it might have been part of a meteor or a small piece of debris that was orbiting the Earth. It was the latter that was becoming a cause for some concern, because the way things were going, the space around the Earth was starting to become a veritable junkyard of discarded and redundant rocket material. A number of technical investigations were also carried out but, as usual, information on all these activities was extremely sketchy. The crew loaded the *Kosmos 1443* spacecraft that had docked with the space station back in March, with some 350 kilogrammes of completed experimental material.

The spacecraft, which was in two sections, would separate after being undocked; one part would return to Earth with completed experiments, whilst the other, which was used as a rubbish bin, would be left to go into a decaying orbit and burn up as it entered the Earth's atmosphere. The crew undocked *Kosmos 1443* from the space station on 14 August and returned it to Earth on 23 August, demonstrating perfect autonomous flight. It soft-landed near Arkaylk, Kazakhstan on 27 August 1983. The remaining part of *Kosmos 1443*, which had separated soon after undocking, went into a decaying orbit and was burnt up as it re-entered the Earth's atmosphere on 19 September.

On 1 and 3 November, the two cosmonauts carried out two EVAs to replace the solar arrays and increase the space station's electrical capacity by 800 watts.

In the meantime *Progress 17* had been launched from Baikonur on a Soyuz-U rocket on 17 August 1983, with more supplies for the crew aboard the *Salyut 7* space station. It was undocked on 17 September with the usual discarded and unwanted materials and placed directly into a decaying orbit to burn up over the Pacific Ocean.

One of the most dramatic moments in the ongoing history of space flight occurred on 26 September 1983. The *Soyuz T-10A* spacecraft, with cosmonauts Gennadi Strekalov and Vladimir Titov aboard, was about to lift off when the rocket exploded. Fortunately for the crew the abort system activated almost instantaneously and the Soyuz capsule was blasted away from the blazing rocket and descended on its parachute a safe distance away. The crew were unhurt and unfazed from the experience, safe in the knowledge that the escape system worked.

Progress 18 arrived on 22 October 1983 after being launched from Baikonur on a Soyuz-U rocket on 20 October. The cargo spacecraft was to stay docked with the *Salyut 7* space station until the arrival of *Soyuz T-10*. One month later, on 23 November 1983, the *Soyuz T-9* spacecraft, with cosmonauts Vladimir Lyakov and Aleksandrov aboard, returned to Earth, landing at Dzhezkazgan, Kazakhstan.

It was to be almost four months before the next crew visited the *Salyut 7* space station. *Soyuz T-10*, with cosmonauts Leonid Kizim, Vladimir Solovyov and Oleg Atkov aboard, blasted off the launch pad on a Soyuz-U rocket at Baikonur on 8 February 1984 and docked with *Salyut 7* two days later. They entered the space station carrying torches, as there was no power. The next few weeks were spent with housekeeping duties, getting the space station habitable and unloading the *Progress 18* cargo spacecraft that had arrived in the previous October. The crew carried out a large number of experiments, including a number of medical experiments by Oleg Atkov who was a doctor. The exact nature of the experiments is not known, as was the normal way with Russia during this period, keeping everything as secret as they could. With the station up and running, Leonid Kizim and Vladimir Solovyov carried out three EVAs to install some external equipment, then later began to carry out repairs on the propulsion system fuel lines, in an attempt to fix a ruptured fuel line.

Another unmanned cargo spacecraft, *Progress 19*, arrived on 21 February 1984 after a two-day flight to the space station. It carried the usual supplies and additional materials and after being unloaded was refilled with unwanted waste. It was undocked on 31 March and placed in a decaying orbit where on 1 April it re-entered the Earth's atmosphere and was burnt up.

Right: Soyuz T-10A exploding on the launch pad.

Below: Inside the cargo spacecraft *Progress 18*. (Roscosmos)

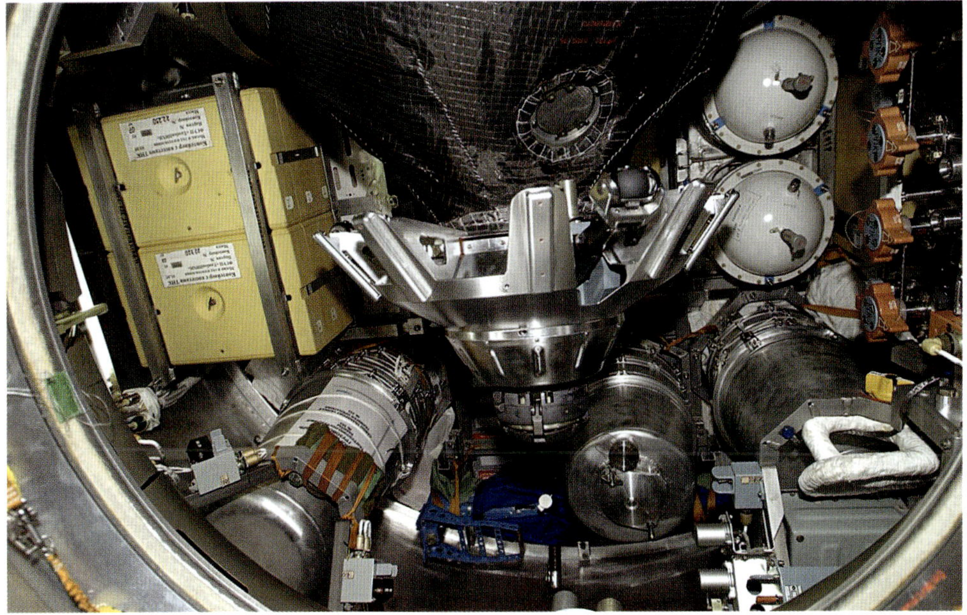

The next Soyuz mission on 3 April 1984, *Soyuz T-11*, carried the first Indian cosmonaut, Rakesh Sharma. The Russian crew aboard *Soyuz T-11* consisted of Yuri Malyshev and Gennadi Strekalov and docked with the *Salyut 7* space station to relieve the crew of Oleg Atkov, Leonid D. Kizim and Vladimir Solovyov. Rakesh Sharma returned to Earth with the crew of *Soyuz T-11* on 11 April.

On 15 April 1984 cargo spacecraft *Progress 20* was launched on a Soyuz-U2 rocket from Baikonur. It contained parts and tools so that the ruptured fuel line for the *Salyut 7* space station propulsion system could be repaired. It was undocked on 17 April and finally burnt up on re-entry on 7 May.

Fresh supplies arrived on the space station on 10 May after a two-day flight from Baikonur on *Progress 21*. The cargo spacecraft also carried a second set of three solar arrays extensions to help increase the power supply. Then on 18 May, the cosmonauts installed the solar arrays to boost the station's power supply. With all the contents unloaded, the cargo spacecraft was refilled with unwanted materials and undocked on 10 May, re-entering the Earth's atmosphere on 26 May. More supplies arrived on 30 May aboard *Progress 22*, which had been launched on a Soyuz-U rocket two days earlier. It was undocked on 15 July and was burnt up on re-entry the following day.

On 17 July 1984 *Soyuz T-12* was launched on a Soyuz-U2 rocket from Baikonur and joined *Soyuz T-11* aboard *Salyut 7*. The crew of Vladimir Dzhanibekov, Igor Volk and Svetlana Savitskaya carried a series of experiments, which included an EVA by Savitskaya who carried out a number of welding experiments outside the space station, the first spacewalk to be carried out by a female. *Soyuz T-12* returned to Earth on 29 July, leaving the Salyut space station empty and floating like a ghost ship in space. Even though the *Salyut 7* space station was now empty, a cargo craft, *Progress 23*, was launched on 14 August 1984 and docked with the space station on 16 August.

The Russians launched their *T-13 Soyuz* spacecraft with cosmonauts Vladimir Dzhanibekov and Viktor Savinykh aboard, on 6 June 1985. This mission was to turn out to be one of the most impressive feats of running repairs in the history of space missions. As they approached the lifeless space station Vladimir Dzhanibekov manually piloted the spacecraft to hard dock with the forward port. The crew checked whether or not the electrical system of *Salyut 7* was alive, but found it dead. They then checked the air within the station before cracking the hatch and found it breathable, but very cold. So donning their winter clothes the two cosmonauts entered the space station. The *Salyut 7* space station had been inactive for almost a year because of a crippled solar array problem and the sight that greeted the two cosmonauts when they entered was a heavy frost covering everything inside. Another cargo spacecraft, *Progress 24*, was launched on 21 June 1985 and docked two days later. It contained replacement parts so that the repair crew could get the *Salyut 7* space station up and running again. It was undocked on 15 July and burnt up on re-entry the same day. This cleared the docking port for the arrival of another unmanned cargo spacecraft, *Kosmos 1669*, which had been launched on 19 July 1985 on a Soyuz-U rocket. It carried more replacement parts enabling the two cosmonauts to achieve one of the most remarkable repair feats in space history. This was the last Progress flight to *Salyut 7* as all attention was on the new *Mir* space station. On 18 September *Soyuz T-14* arrived with Vladimir Vasyutin, Georgi Grechko and Alexsandr Volkov aboard. After a detailed examination of the repairs that had been carried out, Georgi Grechko and Vladimir Dzhanibekov got into the *Soyuz T-13* spacecraft and returned to Earth on 26 September 1985.

The arrival of a modified TKS-4 unmanned spacecraft, *Kosmos 1686*, brought with it a military tracking experiment fitted in a stripped down VA capsule. Amongst the 9,900lb (4,500kg) of freight was a girder that was to be assembled outside of *Salyut 7* together with the crystallization processing apparatus. A major problem arose on 13 November when Vladimir Vasyutin became ill. After lengthy scrambled communications with the TsUP it was decided to cut short the mission and to close the space station down. All three cosmonauts returned to Earth on 21 November, landing in Dzhezkazgan, Kazakhstan. There were conflicting reports as to why the mission was prematurely ended: one was that Vladimir Vasyutin was suffering from a very bad urinary tract infection that was accompanied by a very high fever, the other was that there were psychological issues affecting the mood and performance of his work. The truth will probably never be known, but it has to be remembered that all these missions and situations were being carried out in uncharted territory and could possibly create problems never encountered before.

CHAPTER SEVEN

Mir [Peace]

On 19 February 1986 the core module of the *Mir* (DOS 7) space station was launched on a Proton-K rocket. It provided the living quarters for the cosmonauts along with the environmental systems and the station's main engines. The *Mir* Space Station was the replacement for the Salyut Space Stations and consisted of six docking ports, four at the forward end, one in the aft end and one on the node. The main section, which was 43ft (13.1m) long and 14ft (4.2m) in diameter, was contrived from the *Salyut 6* and 7 space station and contained the command centre, eating area and sleeping area for two cosmonauts. Attached to this was the Kvant module, where the majority of waste was stashed. The Kvant module was 42.6 ft (13.0m) in length and 14ft (4.2m) in diameter. Another of the modules was Kvant 2, which was 44.9ft (13.7m) in length and had a diameter of 14ft (4.3m). This was where the space station's main toilet was housed amongst other things. Also attached was the Kristall module. This 39ft (11.8m) long module had a diameter of 14ft (4.2m) which was in keeping with the other modules. In 1995 an additional module was added – Spektr. This was used primarily by the visiting American astronauts during their stay aboard the space station and contained a variety of scientific equipment. This was followed later by Piroda, a 42ft 6in.(13.0m) long by 14ft 2in (4.4m) diameter module.

The first of the additional modules, Kvant 1, which was attached to a TKS spacecraft, was launched from Baikonur aboard a Proton-K rocket on 31 March 1987. The TKS spacecraft acted as a tug and manoeuvred the Kvant 1 module into position in the aft docking port. With the docking complete, the TKS tug detached itself and went into a decaying orbit around the Earth, eventually burning up on re-entry. This left the docking port empty for the Progress cargo spacecraft to dock and be unloaded. Kvant 1 was equipped with Gyrodines which allowed the space station's attitude to be gyroscopically controlled without using manoeuvring fuel. Inside the module were x-ray and ultraviolet experiments and astronomy observations.

The space station became operational on 6 March 1986 in an orbit just 10 kilometres from *Salyut 7*. Then on 13 March 1986, with cosmonauts Leonid Kizim and Vladimir A. Solovyov aboard, *Soyuz T-15* was launched from Baikonur on a Soyuz-U2 rocket and docked with the space station *Mir* to conduct a series of experiments and investigations. Initially it had been intended that the Soyuz-TM spacecraft would dock with *Mir*'s forward port, leaving the aft port clear for the Progress cargo spacecraft to dock. It was discovered that the older Soyuz-T model spacecraft was not fitted with the Kurs automatic approach system that had been fitted to the forward port on

Mir, but with the older Igla approach system that had been fitted to *Mir*'s aft port. This meant the spacecraft had to approach the aft port of *Mir*, then manually manoeuvre around the space station and dock manually in the forward port after switching off the Igla system. For the next fifty-five days the two cosmonauts carried out tests on the new space station's systems, together with a small number of experiments. With the work completed, the two crew members manoeuvred the space station to the same orbit as that of *Salyut 7*, then on 4 May, caught up with it. The crew then entered the Soyuz spacecraft and undocked from *Mir* and then docked with the *Salyut 7* space station, the first crew ever to visit two space stations on a single mission.

After docking with the *Salyut 7* space station they found it to be totally icebound and without any electrical power. After defrosting the space station, the two cosmonauts carried out an EVA and realigned the solar array on the space station with the Sun in preparation for the manning of the station. After powering up the space station the crew spent the next couple of days making it habitable again and then removed the experiments left behind by the last crew. On 25 June, the crew undocked from the space station for the last time, leaving it to drift in space. The following day *Soyuz T-15* docked again with the space station *Mir*. The *Salyut 7* space

The Russian space station *Mir*. (NASA)

Soyuz TM-2 crew of Yuri Romanenko and Aleksander Laveykin. (Roscosmos)

station was left abandoned and later made an uncontrolled re-entry over Argentina, breaking up as it re-entered the Earth's atmosphere.

On 21 March 1986, fresh supplies were delivered to *Mir* by the unmanned *Progress 25* cargo spacecraft. Launched from Baikonur on a Soyuz-U rocket, this was the first cargo spacecraft to visit *Mir*. After being unloaded, over the next few weeks any unwanted items or rubbish were placed inside and after undocking on 20 April, the unmanned spacecraft was then placed into a decaying orbit and burned up on re-entry over the Pacific Ocean. The crew on *Mir* continued to carry out experiments and investigations on the Sun and other planets in the solar system. A number of medical and biological experiments were also carried out, most of them concerning the effect of prolonged space flight on the human body. Another unmanned *Progress 26* cargo spacecraft docked with *Mir* on 23 April, bringing more supplies, equipment and experiments. Installation work on the inside of the space station continued, and the arrival of another unmanned Soyuz spacecraft, *Soyuz TM-1*, also brought additional supplies and materials. The *TM-1* spacecraft was then used to carry out a number of extensive unmanned flight tests and docking manoeuvres whilst in free space.

On the 28 May 1986 the two cosmonauts, Leonid Kizim and Vladimir Solovyov, carried out an EVA and installed a truss on the exterior of the space station. Two days later they carried out tests on the truss to ensure that it had remained intact. It was decided that the overworked crew had spent long enough in space and on 16 July 1986 they undocked and returned to Earth, landing close to Arkalyak, Kazakhstan.

The now unmanned *Mir's* orbit was raised on 18 January 1987 by using the engine of the *Progress 27* unmanned cargo spacecraft. After docking in the aft port of the space station, the engine on the cargo spacecraft was fired up and the space station's orbit was moved to a higher orbit of 369 kilometres.

On 5 February 1987 *Soyuz TM-2*, with cosmonauts Yuri Romanenko and Aleksander Laveykin aboard, was launched on a Soyuz-U2 rocket from Baikonur. After docking with *Mir* the two cosmonauts powered up the space station making it habitable, then unloaded the docked *Progress 27* cargo spacecraft and then reloaded it with all the waste material before it was undocked from the space station on 23 February and placing it into a decaying orbit to burn up on re-entry. Another Progress unmanned cargo spacecraft, *Progress 28*, docked with *Mir* on 5 March 1987. A number of supplies were transferred, and unwanted refuse was then placed aboard the cargo spacecraft. After undocking from *Mir* on 26 March, the unmanned spacecraft backed away from the space station and deployed a 60m antenna to be used for geophysical experiments. Then, like all the previous cargo spacecraft, it went into a decaying orbit around the Earth and was finally burnt up on re-entry.

The *Mir* space station had its first addition to the modules that made up the station; the Kvant 1 module on 12 April 1987. This was the first addition to the *Mir* complex and was divided into a pressurized laboratory compartment, which was sub-divided into a living area and an instrumentation area and a non-pressurized equipment compartment. Its purpose was to enable cosmonauts to carry out astrophysical observations and material science experiments. Entering the core of *Mir* from Kvant was through a pressurized transfer chamber via the laboratory chamber. The space station received another cargo spacecraft, *Progress 29*, on 21 April with essential supplies for the crew. Two more unmanned cargo spacecraft, *Progress 30* and *Progress 31*, arrived on 13 May and 9 August with more essential supplies. All were used as waste bins and were burnt up on re-entry

Three more cosmonauts on board *Soyuz TM-3* visited the *Mir* space station on 5 August 1987, Alexsander Viktorenko, Aleksandr Aleksandrov and Syrian cosmonaut Mohamed Faris. They were to carry out joint investigations and experiments with cosmonauts Yuri Romanenko and Alexsander Laveykin who were still aboard *Mir*. The *Progress 32* unmanned cargo spacecraft arrived on 23 September with more supplies and replacement parts. Two months later on 20 November the unmanned cargo spacecraft, *Progress 33*, arrived with more supplies and enabled the crew to get rid of their unwanted rubbish, defunct parts and materials.

The arrival of the *Soyuz TM-4* spacecraft on 23 December 1987 brought three more cosmonauts to *Mir*: Musa Manarov, Anatoly Levchenko and Vladimir Titov. Anatoly Levchenko was a prospective pilot for the Buran space shuttle orbiter and the purpose of his brief visit was to introduce him to the space environment that was home to the cosmonauts. Romanenko and Laveykin returned to Earth with Anatoly Levchenko (*Soyuz TM-4*) on 29 December 1987. During the next couple of months the crew completed the solar array installation and carried out a complete exterior inspection of the space station.

On 20 January 1988, *Progress 34* was launched on a Soyuz-U2 rocket from Baikonur with fresh supplies, docking with *Mir* three days later. After being unloaded it was refilled with unwanted materials and waste, undocked and placed in a decaying orbit to burn up on re-entry.

Mir was again visited on 24 March 1988, by the unmanned *Progress 35* spacecraft. After all the supplies and additional experiments and investigations had been delivered, waste material from *Mir* was transferred to the spacecraft. The Progress spacecraft then undocked from *Mir* on 5 May and was burnt up on re-entry. Eight days later another cargo spacecraft arrived, *Progress 36*, which was unloaded and once again filled with unwanted rubbish and then undocked and left to burn up on re-entry.

On 7 June 1988 *Soyuz TM-5*, on a Soyuz-U2 rocket, blasted off the launch pad at Baikonur to rendezvous with the *Mir* space station. On board was a Russian/Bulgarian crew consisting of Russian cosmonauts Aleksandr Solovyev and Viktor Savinykh, and Bulgarian cosmonaut Aleksander Aleksanderov. The crew was to carry out the usual joint investigations and experiments, as well as carrying out the necessary repairs and housekeeping required to keep the space station operational. After seven days the three cosmonauts moved their spacecraft to the aft port to make way for the arrival of the next Progress supply craft.

Another cargo spacecraft, *Progress 37*, arrived on 18 July with more essential supplies. Like all previous Progress cargo spacecraft it was used as a rubbish bin over the next few weeks before being undocked on 12 August and placed in a decaying orbit to be burnt up in the Earth's atmosphere as it re-entered. It is staggering to think that each

Soyuz TM-5 crew of Anatoli Soloviyov, Viktor Savinykh and Aleksandr Aleksandrov. (Roscosmos)

of the Progress cargo spacecraft cost approximately 18 million dollars and were only used the once. The exact figure is not known.

The arrival on *Soyuz TM-6* on 31 August 1988 of Russian cosmonauts Vladimir Lyakhov and Valeri Polyakov, together with a cosmonaut from Afghanistan, Abdul Mohmand, made the conditions aboard *Mir* rather cramped.

On 6 September Lyakhov and Mohmand left the space station aboard *Soyuz TM-5* to return to Earth. A problem arose when, after undocking from *Mir*, the retrofire rockets never activated because of a fault with the infrared horizon sensor, which could not recognize the spacecraft's correct attitude. After waiting seven minutes, Lyakhov tried again, this time the retrorockets fired, but only for three seconds. The reason for this was that Lyakhov did not want a landing overshoot. Three hours later he fired the main engine, and this time it lasted six seconds. Lyakhov then tried to manually put the spacecraft into a de-orbital attitude, but the on-board computer shut the engine down after just six seconds. After three more attempts, the crew was forced to shut down until the spacecraft came into an alignment with the landing area again. This meant that the two crew members had to spend a further twenty-four hours in the cramped descent module with hardly any water or food and no sanitation, before making another attempt. This time it was a successful retrofire and the spacecraft deorbited safely. *Soyuz TM-5* landed near Dzhekzazgan, Kazakhstan on 7 December without further problems. During his mission Abdul Mohmand's mother became so distraught over safety concerns for her son, that President Najibullah of Afghanistan arranged an audio/video meeting between them so as to allay her fears. There was a concerning incident aboard the space station during this mission, when an oxygen generator caught fire. The fire was smothered by Valery Polyakov using a spare flight suit, but the matter was hushed up by the Russian authorities and only emerged when it appeared in a Russian scientific journal a couple of years later. It then emerged, after a number of questions had been asked, that a similar fire had occurred aboard the *Salyut 6* space station in 1978. That fire had been serious and almost caused the evacuation of the space station.

The missions to the space station *Mir* were now becoming a regular pattern in the Russian space program, as were the unmanned cargo spacecraft, and the arrival of *Progress 38* on 9 September showed the need to supply the crews with food and other essentials, as they were staying longer and longer. Cosmonauts from other countries that were financially able and allied to Russia were also being trained and flown aboard the space station, albeit for very short periods. The arrival of *Soyuz TM-7* to *Mir* on 28 November 1988, brought cosmonauts Alexander Volkov, Sergei Krikalev and French cosmonaut Jean-Loup Chrétien. After completing a number of experiments and investigations, the resident crew handed over control of the space station. Cosmonauts Musa Manarov, Vladimir Titov (*Soyuz TM-4*) and Jean-Loup Chrétien returned to Earth on 21 December 1988. During the mission Manarov said that he saw a UFO. Questioned later about the sighting he was adamant that what he saw was definitely not in his imagination. On Christmas Day 1988, the arrival of *Progress 39* no doubt brought some festive cheer to the crew and there were probably some Christmas surprises for them in the cargo.

Because of greater solar activity, the space station was found to have decayed in orbit, so it was decided to use *Progress 39*'s engine to boost *Mir*'s orbit from 325 kilometres by 353 kilometres to 340 kilometres by 376 kilometres. The next unmanned cargo spacecraft, *Progress 40*, was launched on 10 February 1989 on a Soyuz-U2 rocket, with supplies for the *Mir* space station. This was not the normal cargo spacecraft this time, as it was fitted with an ejection seat that was fired as the Soyuz-U2 rocket soared into the atmosphere. The Buran ejection seat had been installed on top of the rocket in place of the escape tower engine and is said to have been a complete success. Also on board the cargo spacecraft was a Bulgarian Spektr 256 spectrometer and power supplies for some of the equipment that had failed on the station. Once docked and the supplies transferred, the Progress engine was fired and used to put *Mir* into a higher orbit of 416 kilometres. Unusually this manoeuvre caused the Progress cargo spacecraft to run out of fuel, which resulted in the cargo spacecraft, when undocked on 3 March, going into an uncontrolled decaying orbit. *Progress 40* entered the Earth's atmosphere two days later and was burnt up on re-entry over the Pacific Ocean.

The flow of Progress unmanned cargo spacecraft continued with the arrival of *Progress 41*. Launched on 16 March 1989 on a Soyuz-U2 rocket, it docked in the aft port of the Kvant module two days later. It was unloaded and refilled with unwanted materials and waste and one week later undocked and placed in a decaying orbit to be burnt up on re-entry. The three remaining crew members in the *Mir* space station, Alexander Volkov, Sergei Krikalev and Valeri Polyakov, having completed all their experiments, returned to Earth on 28 April 1989, landing in Arkaylk, Kazakhstan and leaving the *Mir* space station empty.

The arrival of the new *Progress M-1* cargo spacecraft on 23 August 1989 heralded the start of a new improved version. Launched on a Soyuz-U2 rocket from Baikonur, it docked in the empty space station in the forward port of the *Mir* space station two days later.

The launch of *Soyuz TM-8* on 5 September 1989, with cosmonauts A. S. Viktorenko and Alexander Serebrov aboard, caused the rest of the world to sit up and take notice when the Russian government announced that the flight had cost 80 million roubles. This was the first time that the Russian government had ever mentioned costs and was seen by the West as a relaxing of the barrier that had kept the East and West apart. The spacecraft docked with *Mir* on 8 September and proceeded to carry out a series of technological research programs and investigations that would justify the cost. Whilst they were there, *Mir* was expanded by the addition of the Kvant 2 module. Initially there were problems due to the failure of the automatic rendezvous system and the solar panel, which failed to extend. Fortunately ground control corrected all the faults and Kvant 2 docked successfully with the *Mir* space station on 26 December 1989.

This was the second largest module to be attached to the *Mir* space station and one of the most complex. The module was based on the TKS transport spacecraft that had been used on the Almaz military space station program and did not require a docking arm to guide it to the forward axial docking port on the space station. Kvant 2 was fitted with additional solar arrays, water recycling apparatus, an oxygen generator and an airlock

for EVAs. The module included a second set of gyrodines. The gyrodines, along with those already installed in Kvant 1, provided increased attitude control for the station complex.

Another of the Progress unmanned cargo spacecraft, *Progress M-2*, arrived at *Mir* on 22 December 1989, with supplies and a number of scientific experiments. Among these experiments were some concerned with biotechnology that had come from the United States. The great non-co-operation divide between the two super-powers was being closed up slowly but surely.

On *Mir*, cosmonauts A.S. Viktorenko and Alexander Serebrov installed a star tracker that had earlier been delivered by one of the numerous unmanned Progress cargo spacecraft. They also carried out a detailed inspection of the Kvant 2 module that had been attached to *Mir* as well as a new type of EVA suit, complete with an SPK manoeuvring unit.

On 6 December 1989, during the reconfiguration of *Mir*, the second module that had been added to the complex – Kvant 2 – was moved. Initially it was attached to the forward axial docking port, then on 8 December, it was permanently attached to one of the radial docking ports on the core of the space station. It contained facilities where Earth observation experiments, life and material sciences could be housed, as well as water regeneration plants and additional living quarters. The crew were joined on 12 February 1990 by Vladimir Solovyov and Alexandr Balandin, who had docked in their *Soyuz TM-9* spacecraft after being launched from Baikonur on a Soyuz-FG rocket. They transferred a number of experiments and investigations to the space station and then commenced a six-month program carrying out intensive investigations into geophysical and astrophysical research coupled with experiments on biotechnology and space materials science. Alexandr Viktorenko and Alexander Serebrov returned to Earth aboard the *Soyuz TM-9* spacecraft on 19 February 1990. Another cargo spacecraft, *Progress M-3*, arrived on 28 February with the usual essential supplies.

Another cargo spacecraft, *Progress 42*, was launched on a Soyuz-U2 rocket on 5 May 1990, carrying supplies and equipment in preparation for the arrival of a new piece for the *Mir* space station – Kristall. This third addition to the core vehicle was launched on 31 May 1990, the previous two being Kvant 1 and Kvant 2. The module was scheduled to dock with *Mir* on 6 June, but problems arose with one of Kristall's orientation engines, so there was a delay until a successful docking was achieved on 10 June. The module was divided into two compartments: the instrument-payload compartment and the junction-docking compartment, the latter with a docking module fitted for the Buran orbiter. The instrument/cargo and instrument/docking compartment had a length of 11.9 metres and a diameter of 4.3 metres, and contained a number of furnaces for carrying out crystal growing and conducting experiments for semiconductor production. Two specially designed docking ports were carried, capable of receiving spacecraft weighing 100 tons or more, specifically the Soviet space shuttle orbiter Buran, but it was in fact the American space shuttle orbiters that were the first to use the docking port in 1995. However it was discovered that the American orbiter could not dock on the rear radial ports because there was not enough clearance. To solve the problem the Kristall module was moved to the forward axial port, but this created another problem when it was discovered that Soyuz spacecraft could no longer use the axial ports. A new docking module was designed to alleviate the problem and once installed it would provide the clearance required to enable the orbiters to dock without repositioning the Kristall module every time. It was delivered and fitted by the American Space Shuttle orbiter *STS-74* (*Atlantis OV-104*).

In a move away from the attention of the Soyuz missions to the space station, the Mikoyan-Guerevich (MiG) Company's subsidiary NPO Molniya, together with NPO Energia, unveiled their answer to the defunct Buran Orbiter, the MAKS spaceplane. The spacecraft had a wingspan of 12.5 metres, a length of 19.3 metres and was powered by two RD-701 tripropellant engines, which were fueled by liquid hydrogen and liquid oxygen. It carried a crew of two and a payload of 9 tonnes in the payload bay. To date, mainly because there is no funding readily available, the spacecraft has not flown, but tests are said to have taken place successfully.

The Russian space program continued when *Soyuz TM-10* blasted off the launch pad at Baikonur aboard a Soyuz-U2 rocket on 1 August and docked with the space station *Mir* the following day. On board were cosmonauts Gennadi Manakov and Gennadi Strekalov and they were to join up with fellow cosmonauts Anatoly Solovyov and Aleksandr Balandin to continue the ongoing investigations into the geophysical and astrophysical research experiments. Anatoly Soloviyov and Alexsandr Balandin later returned to Earth in the *Soyuz TM-9* spacecraft on 9 August 1990.

One of the regular Progress cargo spacecraft, *Progress M-4*, arrived on 15 August bringing fresh supplies for the crew. The two remaining cosmonauts continued with their maintenance and housekeeping duties together with their experiments. On 27 September the cargo spacecraft *Progress M-5* docked with the space station and, not only did it bring fresh supplies, but it gave the crew the opportunity to get rid of all their trash by placing it in the spacecraft to be burnt up as the cargo ship entered the Earth's atmosphere.

The two crew members powered up the Kvant 1 module in October 1990, and it was soon back on line showing no ill effects from being shut down for a year. In addition to the scientific section contained in Kvant 1, it also contained gyrostabilizers that had been designed to improve the stability of the space station and reduce the propellant consumption used to keep *Mir* on track. From a safety aspect, Kvant 1 also carried additional life support equipment.

The first commercial paying passenger to be taken into space was launched from the Russian space center at Baikonur on 2 December 1990 aboard *Soyuz-TM-11* on a Soyuz-U2 rocket. He was journalist Toyohiro Akiyama and it cost the Japanese television company, TBS, $28 million for the privilege. Russian cosmonauts Viktor Mikhailovich Afanasyev and Musa Khiromanovich Manarov accompanied him. Akiyama was to join the existing crew aboard the *Mir* space station, Gennadi Manakov and Gennadi Strekalov. The journalist was scheduled to make one ten-minute TV broadcast and two twenty-minute radio broadcasts daily from the space station. It was discovered that the electrical power for his equipment was incompatible with that of the space station, consequently converters had to be sent up in a Progress-M spacecraft. Toyohiro Akiyama's couch was transferred from *Soyuz*

TM-11 to Soyuz *TM-10* in preparation for his return to Earth. On 8 December Gennadi Manakov and Gennadi Strekalov loaded the spacecraft with Akiyama's film and equipment along with some completed experiments. Toyohiro Akiyama returned to Earth aboard *Soyuz TM-10* on 10 December 1990, after spending just eight days in space at the cost of $3.5 million per day, together with cosmonauts Gennadi Manakov and Gennadi Strekalov (*Soyuz TM-10*) who had spent almost 131 days in the space station. The spacecraft landed outside Arkalyk, Kazakhstan, where Toyohiro Akiyama made his final broadcast. Like his fellow Russian cosmonaut Musa Manarov, Gennadi Strekalov also claims to have seen a UFO during his time on *Mir*. Again there were the usual derisory comments, but like Manarov, Strekalov was convinced of what he saw.

The next cargo spacecraft to *Mir*, *Progress M-6*, arrived on 14 January 1991 with much needed supplies in preparation for the resident crew and their replacements later in the year. On 17 March *Progress M-7* cargo spacecraft arrived with fresh supplies and new equipment, enabling the crew to get rid of all their housekeeping rubbish by dumping it all into the now empty cargo spacecraft and sending it into a decreasing orbit to be burnt up as it entered the Earth's atmosphere. *Progress M-7* also carried a VBK-Raduga spacecraft which enabled some of the completed experiments to be returned to Earth; unfortunately it was lost on re-entry.

The second commercial flight to the Russian space station *Mir* took place on 18 May 1991. The crew of *Soyuz TM-12* consisted of Anatoli Artsebarski, Sergei Krikalev and British cosmonaut Helen Sharman. The sum to be paid to the Russians by a British consortium to take Helen Sharman on the flight was around $16 million. A company had been set up, Antequera Ltd., whose job it was to select the cosmonaut and acquire the sponsorship. Well into the training of the two British cosmonauts selected, Helen Sharman and back-up cosmonaut Tim Mace, a major problem arose. Antequera, the consortium that was obtaining the sponsorship for Helen Sharman and Tim Mace, was not able to come up with the balance of the sum required, and had gone into liquidation. The result was that although Helen Sharman was taken to the *Mir* space station, the balance had to be absorbed by the Moscow Narodny Bank. This of course greatly limited the time and number of UK experiments that were carried out during her time aboard the space station. After only eight days in space Helen Sharman returned to Earth in *Soyuz TM-11* on 26 May with cosmonauts Viktor Afanasyev and Musa Manarov, after the two Russian cosmonauts had completed 175 days on the *Mir* space station. They landed near Dzhezkazgan, Kazakhstan on 26 May.

On the 1 June 1991, the cargo spacecraft *Progress M-8* docked in the forward docking port after the *Soyuz TM-12* had been moved to the aft docking port. This was because damage to the locking mechanism on *Mir* meant that Progress's docking mechanism wouldn't work. With all the unwanted rubbish stored aboard *Progress M-8*, the cargo spacecraft was undocked just in time for the arrival of *Progress M-9* on 20 August.

Another visit to *Mir* brought another commercial passenger to the Russian space station. Joining Russian cosmonauts Vladislav Volkov and Toktar Aubakirov, at a cost of $7 million, was Austrian cosmonaut Franz Viehboeck. Their spacecraft *Soyuz TM-13* was launched on 2 October 1991 from the Russian space cosmodrome at Baikonur on a Soyuz-U2 rocket. They joined cosmonauts Anatoli Artsebarsky and Sergei Krikalev 'already on the space station' to carry out a series of experiments and to relieve the existing crew. Franz Viehboeck carried out experiments in electronic measurements during his time on the space station. On 5 October the Austrian Franz Viehboeck, with Russian cosmonauts Anatoli Artsebarsky and Toktar Aubakirov, returned to Earth aboard *Soyuz-T12*. The arrival of *Progress M-10* on 21 October brought much needed supplies and equipment. The first attempt to dock on 19 October had to be aborted, but it managed to dock on the second attempt two days later. The cargo spacecraft carried water, food, oxygen and equipment required for conducting further scientific research. Also attached was a VBK-Raduga capsule which was used to return experimental results and equipment back to Earth when the Progress cargo spacecraft was undocked and deorbited on 20 January 1992. Another of the cargo spacecraft, *Progress M-11*, docked with the *Mir* space station on 27 January carrying supplies, water and oxygen. After being unloaded, it was refilled with waste and unwanted materials. It wasn't until 13 March before it was undocked and placed into a decaying orbit around the Earth.

The commercial side of the Russian space program was helping to provide some desperately needed funds for the scientific side. But some of the countries who had taken up the offer of these flights were beginning to question the value-for-money of such missions, bearing in mind the extremely high costs and the very short time spent on the space station. One of those countries who took up the offer was Germany, and on 17 March 1992, *Soyuz-TM14* on a Soyuz-U2 rocket was launched from Baikonur with cosmonauts Aleksandr Viktorenko, Aleksandr Kalen and German cosmonaut Klaus-Dietrich Flade on board. This was the first Russian space flight since the collapse of the Soviet Union. Klaus Flade carried out a number of German experiments connected with the Freedom and Columbus Space Station project. He returned to Earth on 25 March 1992 aboard *Soyuz-TM13* with Russian cosmonauts Sergei Krikalev and Vladislav Volkov. The two Russians were regarded by some as being the last citizens of the USSR, as they had launched from Kazakh within the USSR and landed in what had become, since their going into space, the Independent Republic of Kazakhstan.

With the fall of Communism came a number of economic problems, and funding for the space program was one of them. Then in February 1992, the Russian Space Agency was established and amalgamated with a number of other smaller organizations. It was later renamed the Federal Space Agency and amalgamated with the Russian Aviation and Space Agency, to form one organization – Roscosmos.

On 19 April 1992 the arrival of another cargo space craft '*Progress M-12*' brought much needed supplies to the space station. It undocked two months later on 27 June and was placed in a decaying orbit. Then in June came a breakthrough in Russian and American relationships, when US President George Bush met with Russian Premier

Boris Yeltsin and announced a pioneering space agreement where one American astronaut would visit the Russian space station *Mir*, whilst two Russian cosmonauts would fly on the Space Shuttle.

The arrival of *Progress M-13* on 4 July brought additional supplies of food and oxygen. Launched on 30 June on a Soyuz-U2 rocket, the first docking attempt on 2 July failed, but locked-on on the second attempt in the forward port. After being unloaded it was refilled with waste and unwanted materials over the next three weeks. It undocked on the 24 July to make the docking port available for the arrival of Soyuz *TM-15* and was placed into a decaying orbit where it entered the Earth's atmosphere and was burnt up over the Pacific Ocean. Four days later, on 28 July 1992, the Russians made another flight to *Mir*. The crew, consisting of Alexsandr Solovyev, Sergey Avdyev and French cosmonaut Michel Ange-Charles Tognini, had blasted off the launch pad at Baikonur on 27 July 1992 aboard *Soyuz-TM15* on a Soyuz-U2 rocket. After carrying out a number of joint experiments with the other two Russian crew members, Alexsandr Solovyov, Sergey Avdeev and French cosmonaut Michel Tognini boarded the spacecraft *Soyuz-TM14* and returned to Earth on 9 August 1992. One week later on 15 August, *Progress M-14* took more supplies to the space station, giving the onboard crew the opportunity to get rid of the accumulated rubbish and unwanted materials. The cargo spacecraft was also modified to transport the first VDU, a self-contained propulsion block system, to *Mir*, along with the sixth VBK-Raduga capsule, which was returned to Earth and recovered later, unlike the Progress spacecraft, which undocked on 21 October 1992 and was placed in a decaying orbit around the Earth. The VDU propulsion unit was mounted on the outside of the space station. Another cargo spacecraft, *Progress M-15*, arrived on 27 October to deliver more supplies, parts and experiments to the space station. Once unloaded it, like all Progress cargo spacecraft, was refilled with waste and unwanted materials, but remained docked until 4 February 1993 when it was undocked. It remained in orbit to take part in the Znamva 2 experiments and research into autonomous flight. It went into deorbit on 7 February and was burnt up over the South Pacific Ocean.

On 24 January 1993 the Soyuz spacecraft *TM-16*, with cosmonauts Gennadi Manakov and Alexsandr Fyodorovich Poleschuk aboard, blasted off the launch pad at Baikonur on a Soyuz-U2 rocket and docked with the *Mir* space station. This was the first Soyuz spacecraft not fitted with a probe and drogue docking system, but an APAS-89 (Androgynous Peripheral Attach System) docking mechanism that replaced the APAS-79. This was originally designed to be used as the docking system for the Buran space shuttle, but was used by *Soyuz TM-16* to dock with an androgynous docking port on the Kristall module in preparation for the docking of the American Space Shuttle Orbiters. Over the next five months a series of Progress unmanned cargo spacecraft visited the space station, bringing necessary supplies and more experiments and investigations. The crew joined up with cosmonauts Solovyov and Avdeed who, after carrying out some of the joint bio-medical experiments that both crews took part in, returned to Earth on 1 February 1993 aboard *Soyuz TM-15*, landing near Dzhezkazgan, Kazakhstan.

The length of time the cosmonauts were spending on the space station was getting longer and longer, with the result that a large number of experiments and investigations were concerned with the psychological and physiological effects of isolation in a weightless environment. This also meant that more and more cargo missions were required to sustain the visiting crews, and on 21 February *Progress M-16* docked in the Kvant 1 module, with much needed supplies of fresh food, water, oxygen and personal items. After being unloaded, the cargo spacecraft was refilled with waste and unwanted items and on 26 March undocked and placed in a decaying orbit to be burnt up on re-entry over the Pacific Ocean. This was followed a month later on 31 March 1993 by *Progress M-17* with more supplies and equipment. It also carried a VBK-Raduga which was later transferred to *Progress M-18*. As with all Progress flights, after being unloaded they were filled over the following weeks, sometimes months, with unwanted materials and rubbish. The cargo spacecraft had originally been scheduled as a normal mission, but remained docked for 132 days whilst docking tests were carried out on the APS-89 system. Because of its extended mission, *Progress M-17* did not have sufficient fuel to enable it to be placed into a controlled deorbit, so on 11 August it was manoeuvred away from the space station and placed in free flight for the next 205 days whilst it gradually deorbited. After reaching a low orbit, using the remaining fuel it was placed into a controlled deorbit where on 3 March 1994 it burnt up as it re-entered the atmosphere over South America.

The Russians continued to maintain the space station *Mir* by keeping it manned by a variety of crews, and *Soyuz TM-17* was no exception. Launched from Baikonur on a Soyuz-U2 rocket on 1 July 1993 with Russian cosmonauts Vasili Tsibilyev, Alexsandr Serebov and French cosmonaut Jean-Pierre Haignere aboard, the spacecraft docked with the space station on 3 July, just twenty minutes after the unmanned Progress cargo spacecraft had undocked. The usual experiments took place between the members of the two crews before Manakov and Poleschuk, together with the French cosmonaut Jean-Pierre Haignere, returned to Earth on *Soyuz TM-16* on 22 July 1993.

In September 1993 US Vice President Al Gore met with the Russian Prime Minister Viktor Chernomyrdin and an agreement was reached that Russia would help the Americans to build an International Space Station (ISS). The US also agreed to pay the Russian Space Agency $400 million and send seven astronauts to live on the space station over a period of time.

Over the next four months three Progress unmanned cargo spacecraft visited the Russian space station, and *Progress M-18*, *M-19* and *M-20* were all burnt up on re-entry after undocking. However, on 3 July 1993, when the *Progress M-18* spacecraft undocked, she released a VBK-Raduga returnable spacecraft, which was used to return a number of successfully completed experiments back to Earth. The remaining section of the spacecraft went into a decaying orbit.

The first Soyuz flight of 1994 to *Mir* took place on 8 January when *Soyuz TM-18* lifted off the launch pad on a Soyuz-U2 rocket at Baikonur. The crew of Viktor Afanasev,

Yuri Usachev and Valeri Polyakov were about to embark on the fifteenth mission to the space station to carry out a 182-day mission. The Soyuz spacecraft docked with the Kvant module on 10 January and the crew transferred to *Mir*. The oncoming crew joined up with the existing crew and carried out a number of bio-medical experiments before cosmonauts Serebrov and Tsibliyev, who had spent almost 197 days on the space station, returned to Earth aboard *Soyuz TM-17* on 14 January 1994. There was an incident just after they had undocked, when Mission Control ordered Tsibliyev to move the spacecraft to within fifteen metres of the Kristall module for a photo shoot of the APAS-89 docking system. Tsibliyev responded by saying that the spacecraft's response was very sluggish and becoming slightly erratic. Serebrov, who was going to take the photographs, warned Tsibliyev that *Soyuz TM-17* was getting too close to one of the solar arrays. On hearing this Viktor Afanasev ordered Yuri Usachov and Valeri Polyakov to get into *Soyuz TM-18* as a precaution in case there was a collision. Serebrov then reported that there had been a collision and with that Mission Control lost all communication. It was discovered that *Mir* had suffered two glancing blows, but no serious damage. The crew inside the space station said that they had felt nothing. Communication was quickly restored. The cause of the impact was found to be an error in a switch in the hand controller in the orbital module, which governed the braking and acceleration and had been switched on. This action disabled the hand controller in the descent module, but fortunately Tsibliyev was able to steer the spacecraft past the space station without any further problems. Two weeks after the incident, on 28 January 1994, another of the unmanned cargo spacecraft, *Progress M-21*, was launched on a Soyuz-U rocket, docking with *Mir* on 30 January. After transferring supplies from the spacecraft to the space station, the Progress spacecraft was refilled with waste and unwanted materials then undocked from *Mir* on 23 March and placed in a decaying orbit to burn up on re-entry.

After many years of playing the game of one-upmanship, 1994 heralded the breakthrough that everyone wanted. This was the first Shuttle flight of 1994 and the first to have a Russian cosmonaut aboard. The addition of Sergei Krikalev had come about because of a co-operation agreement between NASA and the Russian Space Agency in a program on Human Space Flight. It was also the second flight of the SPACEHAB-2 pressurized module and the 100th Get Away Special payload. On 3 February 1994, *STS-60* (*Discovery OV-103*) lifted off the launch pad at the Kennedy Space Center right on time, one of the rare ones. The SPACEHAB module was activated just after the spacecraft *Discovery* (*OV-103*) settled into its orbit and took up one quarter of the space in the cargo bay. It carried 12 experiments, seven life sciences experiments, four material science investigations and a space dust collection experiment.

One of the other payloads in the cargo bay was the Wake Shield Facility (WSF-1). The WSF-1 was an experiment platform that was designed to leave a vacuum wake in a low Earth orbit that was 10,000 times greater than was achievable on Earth. The idea behind the experiment was that defect-free thin film layers of gallium arsenide and other semi-conductor materials could be grown. On the third day an attempt was made to deploy WSF-1 by means of the remote manipulator system arm, but it was abandoned because of radio interference. A further attempt was made the following day, but that too was abandoned after problems were found in the attitude control system. The platform was returned to the cargo bay and the project abandoned for this mission. A series of in-flight medical and radiological experiments were carried out involving both the Americans and the Russian member of the crew. Krikalev also conducted the ongoing Shuttle Amateur Radio Experiment, contacting both American and Russian radio hams. A number of other experiments were conducted from the mid-deck payload, including six Orbital Debris Radar Calibration Spheres, which ranged in size from two to six inches and were designed to help the calibration of radar tracking systems worldwide.

With all the experiments and investigations completed, the Orbiter *Discovery* headed back to Earth. After eight and half days in space and 130 orbits of the Earth the spacecraft touched down at the Kennedy Space Center on 11 February 1994.

Another of the unmanned cargo Progress spacecraft, *Progress M-22*, docked with *Mir* on 24 March 1994. Launched from Baikonur on a Soyuz-U rocket, this was a two-fold mission, the first to deliver supplies and additional experiments and investigations to the space station's crew, and secondly to use its engine to raise the orbit of the space station from 381 x 400 kilometres, to 398 x 399 kilometres. The Progress spacecraft was filled with rubbish and unwanted materials over the next couple of months, then on 23 May it was undocked and placed into a decaying orbit and was burnt up on re-entry over the Pacific Ocean. The following day *Progress M-23* arrived with more supplies and equipment for the resident crew and after unloading suffered the same fate as the previous Progress cargo spacecraft when it re-entered the Earth's atmosphere on 2 July.

On 1 July 1994, Russia launched another manned Soyuz spacecraft, *Soyuz-TM-19*, on a Soyuz-U2 rocket to *Mir*. The spacecraft, with Yuri Malenchenko and Talgat Musabayev on board, docked on 3 July in the aft Kvant 1 module. Both the cosmonauts were on their first space flights and were to have been accompanied by an experienced cosmonaut, Gennadi Strekalov, as this was the rule at the time, but the cancellation of one of two scheduled Progress cargo spacecraft meant that Gennadi Strekalov's couch had to be used to carry much needed supplies. This was only the second time that a rookie crew had been sent into space, the first was in October 1977 on *Soyuz 25*. Yuri Malenchenko and Talgat Musabayev, together with Valeri Polyakov, who had arrived on *Soyuz TM-18*, were to become the next resident crew.

The next cargo spacecraft, *Progress M-24*, arrived on 27 August, but was unable to dock with the Kvant module because of problems with the latching mechanism. A second, third and fourth attempt caused some minor damage to the space station, so it was decided to manually dock the cargo spacecraft in the forward Core port using a previously untried method. Yuri Malenchenko successfully docked the cargo spacecraft, much to the relief of the rest of the crew and Mission Control. Like all Progress cargo spacecraft, after it was unloaded it was refilled with waste and other unwanted materials, then on 4 October 1994, was undocked and placed in a decaying orbit to burn up on re-entry.

Ten days later the cargo spacecraft *Progress M-25* was launched from Baikonur on a Soyuz-U rocket, carrying supplies of food, personal items, oxygen and a number of pieces of equipment for the space station. This time it docked without incident. As usual, once all the supplies had been removed, the spacecraft was then filled up with unwanted items and rubbish. The Progress cargo spacecraft then undocked on

Unmanned Progress cargo spacecraft being launched with supplies for the ISS. (RKS Energia)

Soyuz TM-20 launching from Baikonur. (Roscosmos)

16 February, was placed into a decaying orbit and left to burn up on re-entry, leaving the aft docking port free for the arrival of *Soyuz TM-20*. Later Yuri Malenchenko and Talgat Musabayev carried out an EVA to inspect the damage caused by the collision of *Soyuz TM-18*, which fortunately turned out to be relatively minor. They carried out a second EVA on 27 September to inspect the solar panels and girder on the Kvant module, but no problems were found. They also replaced a number of minor parts and bolts that had deteriorated or had come loose.

With the work completed the resident crew on *Mir* were joined on 6 October 1994 by the arrival of *Soyuz TM-20* with Alexandr Viktorenko, Yelena Kondakova and European Space Agency astronaut Ulf Merbold aboard. Yelena Kondakova was the third female cosmonaut to go into space and the first to take part in a long distance flight. Ulf Merbold 'on the other hand' had been in space twice before on US Space Shuttle missions, so acclimatized to being in space very quickly. Concerns about the docking latches still worried the crew of the space station, so on 3 November they decided to test the Kure Automatic Approach System (KAAS). Musabayev, Malenchenko and Merbold clambered into *Soyuz TM-19* and undocked from *Mir*. They backed 190 metres away from *Mir* then activated the Kure system and successfully re-docked the spacecraft and transferred back aboard. The German cosmonaut, Ulf Merbold, had been placed on *Mir* to conduct the Euronir 94 experiment for the ESA, which included twenty-three life sciences, four material sciences and three technology sciences. But it had limitations, because the ESA had only budgeted $60 million, forcing Ulf Merbold to rely heavily on the equipment left

The crew of *TM-20* on board the space station *Mir*. (Roscosmos)

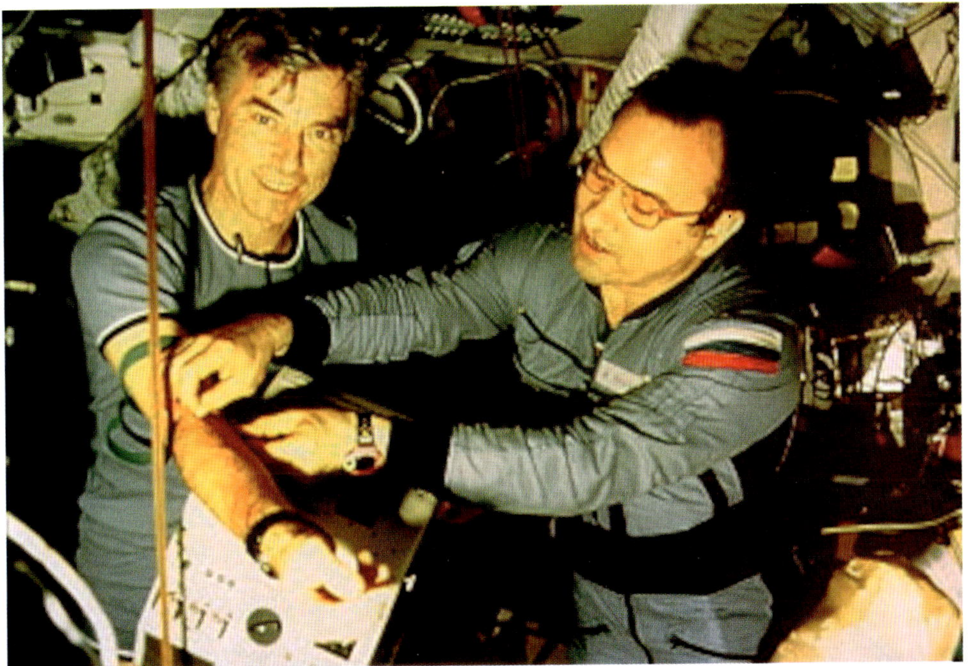

Aleksandr Viktorenko carrying out a medical check on fellow cosmonaut Ulf Merbold whilst aboard the space station. (Roscosmos)

behind by former European cosmonauts, which included the Czechoslovakian-built CSK-1 material processing furnace.

After completing their medical, biological and other investigations and experiments, the cosmonauts Viktor Afanasev and Yuri Usachev climbed into the *Soyuz TM-18* spacecraft and returned to Earth on 9 July, landing north of Arkalyk, Kazakhstan.

The Russians had been carrying out experiments with a new type of Orbiter spacecraft, the Buran, over the past couple of years – all of the tests had been unmanned. Then on 1 December 1994 it was decided that they would carry out the first manned flight. The Buran spacecraft, developed by NPO Energia, was fitted with ejection seats and life support systems and would make its first manned space flight to *Mir*. But for the past few years there had been arguments between the manufacturers and the Soviet Air Force about who would fly the spacecraft. The manufacturers wanted the two-man crew to consist of a pilot and a flight engineer – whilst the Soviet Air Force wanted both crewmen to be qualified military test pilots. Such was the ferocity of the arguments that the Buran project was in danger of being cancelled, so when funding was suddenly withdrawn on 30 June 1993, Russian Premier Boris Yeltsin had the Buran program cancelled. There had only been one flight of the Buran and that was unmanned. Later a universal docking adapter, which had originally been developed for the Buran, was delivered to the space station and which enabled the American Space Shuttle orbiter to dock with *Mir*. Today the Buran spacecraft is just a restaurant attraction in a Moscow theme park, an ignominious end to a project that promised a great deal, but achieved nothing.

On 3 February 1995 *STS-63* (*Discovery OV103*) with James D. Wetherbee (Commander), Eileen M. Collins (Pilot), Mission Specialists Bernard A. Harris, Jr., Michael Foale, Janice Voss and Vladimir G. Titov (R) aboard, blasted off the launch pad and into orbit around the Earth. It had been scheduled for launch on 2 February, but this was postponed until the following day when one of the three inertial measurement units on *Discovery* failed.

The first day was taken up by a series of manoeuvres to bring the spacecraft into the same orbit as that of *Mir*. The crew of the Orbiter *Discovery* contacted the crew aboard *Mir* requesting permission to close within thirty-two feet. With permission given, the

Valeri Polyakov watching the *STS-63* Space Shuttle *Discovery* approach the Russian space station *Mir* in February 1995. (NASA)

Orbiter Discovery closed and maintained station alongside *Mir*. Vladimir Titov, who had spent more than a year aboard *Mir*, contacted the three Russian cosmonauts, Alexander Viktorenko, Yelena Kondakova and Valeri Polyakov, the latter having just broken Vladimir Titov's space endurance record. But this wasn't just a social visit to the Russian space station, it was also the first flight in a series of flights leading to the International Space Station (ISS) program. With the social pleasantries completed, *Discovery* moved away from *Mir* and carried out a fly-around of the space station, filming and photographing it from every angle. They also took a number of shots of the Earth, including a dramatic nighttime shot of the USA.

In the orbiter's payload bay was SPACEHAB-3 which was carrying eleven bio-technology experiments, three advanced materials investigations and six other experiments. In addition to these experiments were the Astroculture experiments, in which plants were grown in a microgravity environment. One important pharmaceutical experiment known as Immune exploited the known tendency of spaceflight to suppress the immune system. This enabled medical scientists to carry out tests on a variety of substances that might prevent or reduce suppression so that people suffering immunosuppressant diseases such as AIDS could benefit. The Remote Manipulator System (RMS) arm was put to good use when the *SPARTAN-204* satellite was lifted out of the payload bay and placed in a free-flying orbit for forty hours. During this time its Far Ultraviolet Imaging Spectrograph instruments carried out investigations on celestial targets and the gas and dust which filled the space between the stars. When the time came to recover *SPARTAN*, two astronauts, British born Michael Foale (A) and Bernard Harris (A), performed an EVA on the end of the RMS arm to carry out trials in handling satellites and assembling techniques. In addition they carried out tests on their modified spacesuits to keep them warm during EVAs at night, but despite the modifications, both astronauts still complained of being extremely cold during this part of the investigation. A number of other ongoing experiments were in the payload bay: The US Air Force Maui Optical Site, the IMAX camera; the Solid Surface Combustion Experiment – on board for the eighth time. After over eight days in space and 129 orbits of the Earth, the Orbiter *Discovery* touched down at the Kennedy Space Center on 11 February 1995.

Another of the Progress unmanned spacecraft, *Progress M-26*, was launched on a Soyuz-U rocket on 15 February 1995 to rendezvous with *Mir*. The spacecraft docked on 17 February with a number of stores and experiments, including oxygen and water. After transferring the stores to the space station the spacecraft was, like all the other cargo spacecraft, filled with rubbish then undocked on 15 March and placed into a decaying orbit and left to burnt up on re-entry.

Another milestone in the ongoing relationship between Russia and America was reached when on 14 March 1995 *Soyuz TM-21* was launched from Baikonur on a Soyuz-U2 rocket. What made this mission special, was that beside the two Russians, Vladimir Dezhurov and Gennadi Strekalov, the other crew-member was American astronaut Norman Thagard – the first American to be carried in a Russian spacecraft. The spacecraft docked with *Mir* on 16 March and Norman Thagard joined

Members of the crew of the *STS-63* orbiter *Discovery* looking up at the Russian space station *Mir* in February 1995. (NASA)

The United States of America at night taken from the ISS. (NASA)

the crew of the space station to carry out a number of experiments and investigations. On 22 March Alexsandr Viktorenko, Yelena Kondakova and Valeri Polyakov said their farewells, climbed into the *Soyuz TM-20* spacecraft, undocked from the space station and returned to Earth, landing safely outside Arkalyk, Kazakhstan. When the spacecraft landed, Valeri Polyakov had spent 438 days in space on one mission, becoming the world record holder.

On the 9 April 1995, the *Progress M-27* cargo spacecraft was launched from Baikonur on a Soyuz-U rocket, docking on 11 April in the forward port of the Core module and bringing much needed supplies to *Mir*. Despite the American Space Shuttle now being available, the Progress cargo spacecraft were still the prime movers of supplies and other necessary equipment and would be for some time to come. They were also the depositories for unwanted materials and waste, which when loaded into the *Progress M-27* spacecraft, were put into a decaying Earth orbit on 22 May to be burnt up on re-entry into the atmosphere over the Pacific Ocean.

The 69th Shuttle flight (*STS-71*), and the 100th space launch with human beings aboard, was also to be the first US spacecraft since the Apollo/Soyuz mission to dock in space with a Russian spacecraft – the space station *Mir*. The main object of the mission was to carry out the exchange of space station crews. The ongoing crew consisted of Anatoly Y. Solovyev and Nikoal M. Budarin. The Russian crew returning from *Mir* were Vladimir Dezhurov, Gennadi Strekalov and American

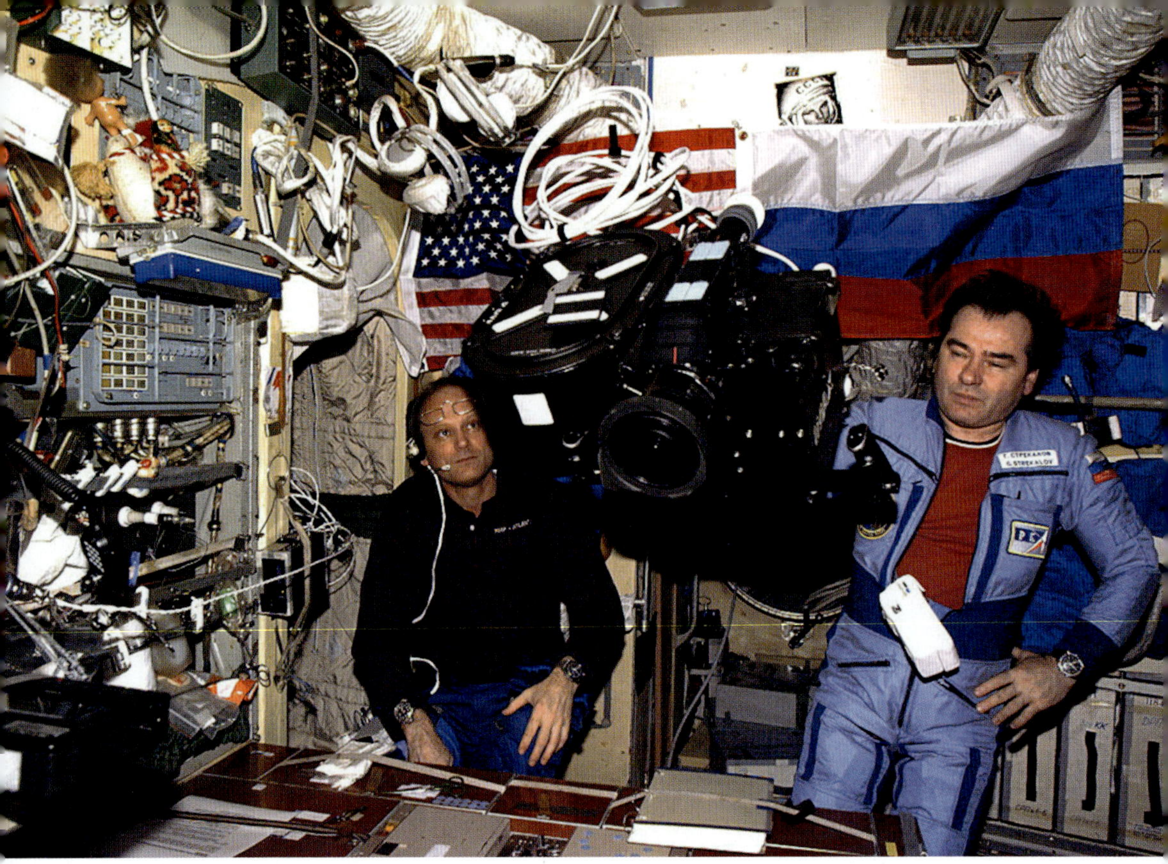

Vladimir Dazhurov and Gennadi Strekalov inside *Mir* after being delivered by the space shuttle *STS-71*. (NASA)

Norman E. Thagard. The launch had originally been planned for the last week in May, but had been delayed to accommodate the Russian activities that were necessary for the docking of the two spacecraft. Included in the activities were a series of spacewalks and the launch of the Spektr module that contained US research hardware. Launched on a Proton rocket from Baikonur on 1 June 1995, the 19-ton Spektr module measured 12 metres by 4.5 metres and was equipped with four solar arrays. Its main purpose was to observe Earth's natural resources and its atmosphere. In addition to this it was to be used to conduct research into a number of space technologies, biotechnology and life sciences. It was also designed to be used as the living quarters for American astronauts.

A new date of 23 June was set for the launch of *STS-71*, but this time it was the weather that intervened and prevented the fuelling of the Space Shuttle's external tank, so the date was set for the following day. At T-9 minutes the weather interfered again and the launch was aborted until the 27th. On 27 June 1995 the Space Shuttle *Atlantis* blasted off the pad at the Kennedy Space Center into space and into the history books. Once in orbit the commander, Robert 'Hoot' Gibson, set about setting the spacecraft in the right trajectory to effect the docking. It was decided that a manual docking approach was the best way to carry out the complicated manoeuvre. Arriving at 250 feet below the space station, Gibson put the Orbiter into a station keeping orbit and awaited further instructions and confirmation from both the United States and Russian mission controllers to continue and dock. Confirmation received, the spacecraft was guided at the rate of 0.1 feet per second: the closing velocity of the two spacecraft was a mere 0.0107 feet per second. On contact there was less than one inch lateral misalignment and an angular misalignment of less than 0.5 degrees per axis and the Orbital Docking System, located in the forward payload bay, performed flawlessly during the docking sequence. The docking took place 216 miles above the Lake Baykal region of the Russian Federation. This was the perfect example of what was meant by space co-operation.

Orbiting at 218 miles above the Earth, *Atlantis* and *Mir* formed the largest spacecraft ever in orbit. The two hatches were opened and the two crews greeted each other – the crew of the *STS-71* then entered the *Mir* space station for the official welcoming ceremony. The formalities over, the two Russian crews exchanged places – the *Soyuz 18* crew transferring to the American spacecraft.

For the next five days a series of joint experiments and programs were carried out, including bio-medical investigations using the Spacelab that was stored in the aft section of *Atlantis*'s cargo bay. Seven different investigations were covered: human metabolism, hygiene, sanitation, radiation, biology, cardiovascular and pulmonary and neuroscience. Medical research was also carried out on the three members of *Mir* who were returning to Earth, whilst they performed a program of exercise in preparation for their return to Earth and its gravity. These experiments were vitally necessary as the long-term effects of weightlessness on the human body was still an unknown quantity. Numerous medical samples including saliva, urine, blood, water and breath were transferred from *Mir* to *Atlantis*. Over 1,000 pounds of water that had been generated by the Orbiter for general purposes was transferred to *Mir* from *Atlantis* along with some nitrogen and oxygen from the Shuttle's environmental control system.

On 4 July, after a farewell ceremony, the two spacecraft undocked and the Orbiter *Atlantis* moved away into a station keeping orbit. A number of continuing experiments were carried out during this time – a number of which were to help the three returning *Mir* crewmen go through an intensive program of exercise after their three months in weightless space. On 7 July, *Atlantis* prepared to re-enter the Earth's atmosphere. On board the spacecraft the three returning members of the *Mir* 18 crew lay in a supine position in custom-made Russian seats that had been installed in the Orbiter's mid-deck. The spacecraft *Atlantis* touched down at the Kennedy Space Center after almost ten days in space and 153 orbits of the Earth.

Another of the Russian cargo spacecraft, *Progress M-28*, was launched from Baikonur on a Soyuz-U rocket on 20 July 1995 taking much needed supplies to the resident *Mir* crew. After docking in the aft docking port on 22 July, the cargo spacecraft was filled with rubbish and unwanted materials and left docked until 4 September when it was released and sent into a decaying Earth orbit to be burnt up on re-entry over the South Pacific Ocean.

STS-71 docked with the *Mir* Space Station. (Roscosmos)

The Russians launched another of their Soyuz spacecraft, *Soyuz TM-22*, on a Soyuz-U2 rocket to *Mir* on 3 September 1995, docking in the forward port. The crew was a real mixture, consisting of Colonel Yuriy Pavlovich Gidzenko, Sergey Vasileyvich Avdeyev and German cosmonaut-researcher Thomas Reiter of the European Space Agency (ESA). The now crowded space station needed more supplies and on 8 October, *Progress M-29* was launched on a Soyuz-U rocket. The cargo spacecraft docked in the aft port of the Kvant 1 module two days later. After being unloaded it was refilled with the usual waste and unwanted materials and on 19 December was placed into a decaying orbit where it was burnt up over the Pacific Ocean.

Another milestone was reached on 17 November 1995, when the Space Shuttle Atlantis *STS-74* (*Atlantis OV-104*) delivered the Kristall Docking Module (DM) to *Mir*. This was the fourth mission of the Space Shuttle to *Mir* and the second docking

The Russian space station *Mir* with the Earth below. (NASA)

mission. It was also the first visit to *Mir* by a Canadian astronaut, Chris Hadfield. *Atlantis* carried the Russian-built Docking Module that was to be attached to the Kristall module on *Mir*. The DM was constructed of an aluminium alloy and covered by Screen Vacuum Thermal Insulation (SVTI) and a micrometeoroid shield. Fixed to the DM was a truss structure that allowed the Shuttle to latch on whilst it was horizontal in the cargo bay. On release, the truss remained attached to the Kristall module and the DM was released. Two solar arrays were also delivered, one American and one that was a joint development. *Atlantis* undocked from *Mir* on 18 November by means of springs that gently pushed the shuttle away from the space station before it fired its jets. *STS-74* landed back at the Kennedy Space Center (KSC) two days later. On 18 December 1995 *Progress M-30* was launched from Baikonur on a Soyuz-U rocket taking much needed supplies to the *Mir* space station and docking in the aft port. *Progress M-30*, re-filled with unwanted materials and waste, undocked on 22 February 1996 and was burnt up on re-entry over the Pacific Ocean the same day.

On 21 February 1996 *Soyuz TM-23* on a Soyuz-U rocket, blasted off the launch pad at Baikonur Cosmodrome, and two days later docked with *Mir*. On board were two Russian cosmonauts, Yuri Onufrienko and Yuri Usachyov, who joined the resident crew of Yuriy Pavlovich Gidzenko, Sergey Vasileyvich Avdeyev and Thomas Reiter.

Whilst on board they carried out a series of investigations and experiments beside carrying out the normal running repairs, servicing and housekeeping required to keep the space station active. They also installed and deployed an MCSA (*Mir* Co-operative Solar Array) that had been delivered by *STS-74*.

After spending 179 days in space, the crew of *Soyuz TM-22*, Yuriy Pavlovich Gidzenko, Sergey Vasileyvich Avdeyev and Thomas Reiter, disembarked from *Mir* in *Soyuz TM-22* on 29 February 1996 and returned to Earth, landing near Arkalyk, Kazakhstan.

Another visit by the American Space Shuttle *STS-76* (*Atlantis OV-104*) was made on 24 March 1996 and after docking, transferred American astronaut Shannon Lucid who was to become a member of the *Mir* crew. She was the first American woman to live on the space station, joining the two Russian cosmonauts, neither of whom spoke English. Despite this problem and Shannon Lucid's meagre grasp of the Russian language, with the continuing help of her two crew members, they managed to interpret and understand each other to the extent of using a Russian/English dictionary just the once during the entire mission. The mission itself was regarded as one of the most harmonious of all

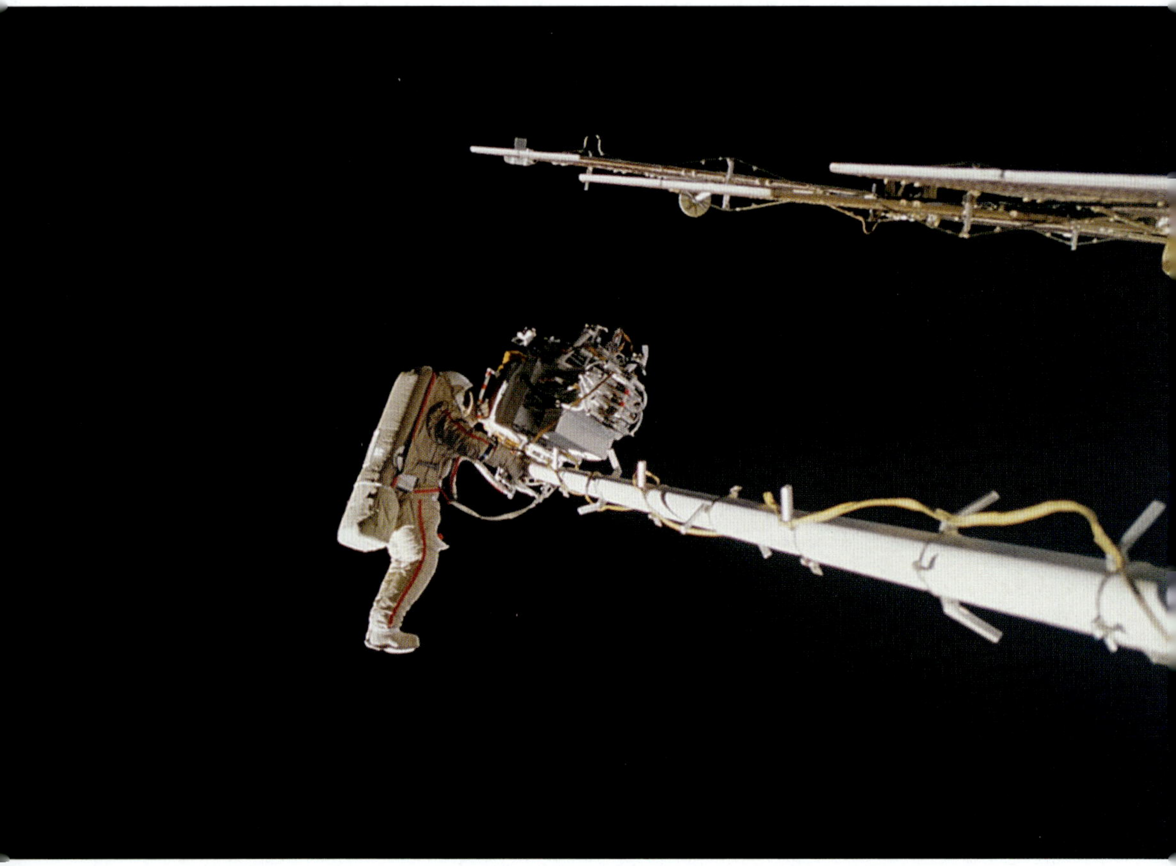

Russian cosmonaut Yuri Onufrienko carrying out a repair on *Mir* during his EVA. (NASA)

the international missions to *Mir*. Although the experiments were carried out individually, there were some joint portions of the mission. One of these was the unloading of supplies and experiments for both cosmonauts and astronauts, including one ton of science equipment and supplies for the US and almost two tons of Russian supplies and experiments that had been brought by *Atlantis*. One and half tons of water, contained in fifteen containers, were also unloaded, as well as three Russian storage batteries that had been serviced and recharged. In return for the delivery, almost one ton of waste, excess equipment and scientific results were placed aboard the *Atlantis* to be returned to Earth. One of the specialist hardware items taken to *Mir* was a seat liner kit for American astronaut Shannon Lucid, which was designed to be placed in a Soyuz module in case of an emergency extraction from the space station. On 28 March *Atlantis* undocked from *Mir* and returned to Earth. A number of research and scientific experiments were carried out, including one that used a high temperature melting oven called the Optizon. The final part of the *Mir* space station, Piroda, arrived on 26 April 1996 after being launched from Baikonur on a Proton-K rocket and was to be used primarily to conduct Earth resource experiments through remote sensing.

The station was visited on 5 May by one of the unmanned Progress cargo spacecraft, *Progress M-31*. After docking in the forward port of the Core module, the crew removed its cargo of supplies, personal items and materials. It was then filled with unwanted rubbish over the next couple of months, then undocked on 1 August and put into a decaying Earth orbit where it burnt up on re-entry. *Progress M-32* arrived on 31 July to re-supply the crew on the space station, docking in the aft port of the Kvant 1 module, but it wasn't until 20 November that it was undocked and placed in a decaying orbit around the Earth.

The continuing expeditions to the *Mir* space station brought another international crew when *Soyuz TM-24* was launched on 17 August 1996. The crew consisted of two Russians, Valeri Korzun and Alexsandr Kaleri, and one female French cosmonaut, Claudie André-Deshays (Haignere), who was the wife of French cosmonaut Jean-Pierre Haignere and was only to stay on the station for a matter of two weeks before she returned with the crew of *Soyuz TM-23*. On 2 September 1996, after spending 193 days in space, Yuri Onufrienko, Yuri Usachev and Claudie André-Deshays, who had spent just fourteen days in space, returned to Earth on board the *Soyuz TM-23* spacecraft. The spacecraft landed 108 kilometres south-west of Tselinograd.

The arrival of the relief crew member, John Blaha, on the Space Shuttle *STS-79* (*Atlantis OV-104*), to the Russian space station on 19 September 1996, was a great relief to American astronaut Shannon Lucid who had spent the last six months on *Mir*. Shannon Lucid returned to Earth aboard the *STS-79* and landed at KSC on 26 September. During her stay aboard the Russian space station, Shannon Lucid followed a rigid exercise routine, which, when after landing at the Kennedy Space Center, enabled her to walk off *Atlantis* unaided. A large number of supplies were also delivered to the resident crew aboard the space station.

Above: Yuri Onufrienko carrying out maintenance on *Mir*. (NASA)

Below: Yuri Usachov carrying out maintenance on one of the cable booms on the ISS. (NASA)

The next cargo spacecraft, *Progress M-33*, was launched on 19 November 1996, arriving just after *Progress M-32* had left, leaving the aft port of the Kvant 1 module clear. After being unloaded and refilled with the usual waste, it was undocked on 6 February in a re-docking exercise. An attempt to re-dock on 4 March failed and the cargo spacecraft was left in orbit until 12 March 1997 when it was placed in a decaying orbit to be burnt up on re-entry.

The next visit by an American was on 12 January 1997 when *STS-81 Atlantis* docked with *Mir*, and American *Mir* crew member John Blaha was replaced by US Navy flight surgeon Jerry Linenger. One month later, on 12 February, he was joined by the next Russian crew to arrive. *Soyuz TM-25* docked with *Mir* with Russian cosmonauts Vasili Tsibliev and Alexsandr Lazutkin, and German cosmonaut Reinhold Ewald aboard. This was the second time a German cosmonaut had gone into space with the Russians and it was rumored that the German government paid $60 million for the privilege.

The crew of *Soyuz TM-25* had replaced Korzun and Kaleri and less than one month later, problems on the *Mir* space station began. The only way to describe this mission was a series of problems, mishaps, fires, collisions and breakdowns, starting when a fire broke out in one of the oxygen generating lithium percolate cartridges in the Kvant module. Alexsandr Lazutkin had just replaced the cartridge when he heard a hissing sound, then he saw some sparks which were followed by a flash of flame. The fire was intense and lasted fifteen minutes, shooting flames a metre long. The incident happened in an area between the crew and the two Soyuz spacecraft that were attached, preventing the crew getting to them had the fire got out of hand. The Russian crew members fought the blaze with extinguishers and managed to put the fire out, but such was the intensity of the smoke, that the crew had to wear gas masks for two and half hours afterwards. One week later the carbon dioxide removal system failed and it was to be some days before the next cargo spacecraft was due to arrive. The cargo spacecraft, *Progress M-34*, was launched from Baikonur on a Soyuz-U carrier rocket and arrived on 6 April 1997 with much needed supplies which included two new spacesuits, three fire extinguishers and repair equipment for the carbon dioxide removal system.

On 2 March 1997, after eight months in space, the *TM-24* crew, Valeriy Korzun and Alexsandr Kaleri, returned to Earth with German cosmonaut Reinhold Ewald, who had joined the space station one month earlier on 7 February. Their spacecraft landed in Dzhezkazgan, Kazakhstan.

The American Space Shuttle *STS-84* (*Atlantis OV-104*) docked with *Mir* on 15 May 1997, with replacement crew member Michael Foale, who was to replace Jerry Linenger and join the Russian crew for the next four months. This was the sixth of nine planned missions to the Russian Space Station and Astronaut Jerry Linenger returned with *STS-84*. Then on 25 June 1997, Vasili Tsibliev, using a newly developed remote-control system, undocked the *Progress M-34* cargo spacecraft, which was full of unwanted rubbish and materials, in a procedure to test the manually-operated docking system. This was an undock and redock test designed to establish whether or not the automatic docking system could be eliminated in an effort to reduce the ever spiralling costs. As the spacecraft approached the space station to re-dock, it went off course and collided firstly with one of the solar arrays on the Spektr module and then with the module itself. A large hole was punched in the solar array and one of the radiators was damaged, but the really serious damage was caused to the Spektr module itself, when it was discovered that a hole had been punched through the skin. A hissing sound was then heard and the module began to depressurize. It was made even worse because the cargo spacecraft was in the process of being docked manually and the station was being depressurized at the same time. The hatch to the now leaking Spektr module was sealed leaving all Michael Foale's personal effects and a number of NASA experiments inside. The remainder of the station was then re-pressurized. The crew closed the hatch only to find that the electrical connection between the Spektr's solar panels and the main station had been severed in the accident. Following the incident, the *Progress M-34* cargo spacecraft was placed into a decaying orbit, but so concerned were the crew, that Michael Foale had been ordered by the commander, Vasily Tsibiliev, to make ready the Soyuz 'lifeboat' spacecraft and was in it when the unmanned Progress sped by. The cargo spacecraft then went into a decaying orbit where it burnt up on re-entering the Earth's atmosphere. With the emergency deemed to be over, Vasily Tsibiliev and Alexsandr Lazutkin then prepared to carry out a space walk to re-connect power cables to three remaining undamaged solar arrays, but this had to be abandoned when it was discovered that Vasily Tsibiliev had an irregular heartbeat during a routine medical examination. Michael Foale then began training for the space walk, but during one of the training exercises there was a complete power failure in the space station. It was discovered later that one of the main power cables had been inadvertently disconnected. The fault was quickly discovered and the power was restored.

It was decided to wait until the new *Soyuz TM-26* crew arrived before another space walk was to be attempted. With this in mind NASA decided to swap Wendy Lawrence, scheduled to take over from Michael Foale, with David Wolfe, simply because she would not fit into the Orlan-MK EVA suit used by the Russians, and David Wolfe could. In the meantime the space station ran on back-up power. Michael Foale complained that after the incident Vasily Tsibiliev made excuse after excuse in an attempt to stop him using the radio and relaying the problems they were having back to NASA. It wasn't until much later that NASA learned of the incident, which was typical of the secretive attitude still being adopted by the Russians.

Since the fall of communism in Russia in 1992 and the break-up of the Soviet Union, which created almost insurmountable problems with the loss of Ukraine, the war in Chechnya and the rocketing cost of living, the space program found itself at the bottom of the financial pile. With the long cold winter about to descend on Russia and the economy in shambles, outside of Moscow poverty was rife, factories were closed and left abandoned, while workers, including scientists, doctors, teachers, nurses, police officers and even the military, had not been paid for months. Russians were not prepared for the new free economy they had been thrown into and were struggling. The head of the Russian Space Agency, Yuri Koptev, appeared in front of

Boris Yeltsin and the Committee in Moscow and told them, in no uncertain terms, that unless there was a huge injection of money into the space program, it was going to collapse. From being the most dominant leader in space development, Russia was now lagging way behind. As one example, the Russian Space Agency said they could no longer afford to buy the expensive docking systems that they had been purchasing from the Ukraine.

The agreement between the United States and Russia to build an International Space Station (ISS) also came into crisis in April 1997, when NASA representatives in Moscow reported back that the Service Module, which was to be the core of the space station, was nowhere near ready for operational use. The Service Module was essential to the success of the project, because without it the remaining parts of the station would not have the rocket power to re-boost the station into orbit. The Clinton administration issued an ultimatum to the Russian Premier Boris Yeltsin, which resulted in Yeltsin plunging his country ever deeper into debt, but it saved the project.

On 5 July 1997, another of the unmanned Progress cargo spacecraft, *Progress M-35*, was launched from Baikonur on a Soyuz-U2 carrier rocket, which then docked with the aft docking port on the *Mir* Kvant module. The cargo spacecraft was carrying repair equipment for *Mir*'s oxygen generators and a couple of new EVA spacesuits along with the usual food and water supplies. It also carried a 'care' package of personal items for Michael Foale because all he had was a toothbrush given to him by Vasily Tsibiliev. After the supplies had been transferred, the main engine of the Progress spacecraft was used to move the space station to a higher orbit. The spacecraft then undocked from *Mir* and re-docked in the forward port to make way for the arrival of *Soyuz TM-26*. It also carried out re-docking tests using a newly developed remote-controlled procedure that took the place of the automatic system provided by a Ukrainian company.

On 5 August 1997 *Soyuz TM-26* lifted off the launch pad at Baikonur Cosmodrome on a Soyuz-U rocket, to take cosmonauts Anatoly Solovyev and Pavel Vinogradov to *Mir*. It was with great relief that the resident crew handed the space station over to the two replacement cosmonauts, who were to become the next resident crew to live there. Once aboard the station the new crew, together with Vasily Tsibiliev, Alexsanr Lazutkin and Michael Foale, set about carrying out the necessary repairs and maintenance that was required to keep *Mir* 'alive', as well as the normal day to day housekeeping duties. They then carried out an EVA and re-connected the power cables going to the Spektr module and carried out a detailed examination of the module itself. The following day they carried out an IVA (Internal Vehicular Activity), during which time they recovered Michael Foale's equipment and personal belongings that had been left behind at the time of the accident. Despite carrying out a detailed examination of the Spektr module, the hole in the skin was never discovered. During the mission Vasiliy Tsibiliev took part in a television advertisement promoting a brand of UHT milk for a company called Tnuva's. This was the first advert ever to be filmed in space and, as far as is known, the only one.

On 14 August the *TM-25* crew of Vasiliy Tsibiliev and Aleksandr Lazutkin entered their Soyuz spacecraft and headed back to Earth, leaving Michael Foale on the space station with the crew of *TM-26*. As they approached touchdown the retro landing rockets failed to ignite, with the result that, with only the parachute to slow them down, they experienced one of the roughest landings of any returning cosmonaut crew, but fortunately neither suffered any injuries. The spacecraft landed close to Arkalyk, Kazakhstan.

Back on *Mir*, Michael Foale and Anatoly Solovyev carried out a six-hour EVA on 5 September to assess the damage caused by the collision and pinpoint the problem area, but found nothing.

Then on 25 September 1997 the skies around the Kennedy Space Center lit up as *STS-86* (*Atlantis OV-104*) lifted off Launch Pad A and into orbit around the Earth. Amongst the crew was Mission Specialist Vladimir Georgievich Titov (R) and he was to stay aboard *Atlantis* whilst the *Mir* crew transfer of Mission Specialist David Wolfe replacing Michael Foale took place. Two days later on 27 September *Atlantis* arrived at *Mir* and the exchange between Michael Foale and David Wolfe took place.

The launch of *STS-86* had been held up not because of any technical problems or adverse weather conditions, but for final permission to be granted for the docking with *Mir*. The reason for the reticence of the NASA hierarchy was the recent problems the Russians had been experiencing on the space station – such as fires and the breaking down of equipment. This was the seventh space shuttle mission to dock with *Mir* and Astronaut David Wolfe became the sixth US astronaut to become a member of the space station's resident crew. A joint US/Russian EVA then took place when Scott Parazynski and Vladimir Titov fitted a 121-pound Solar Array Cap to the docking module to enable future *Mir* crew members to seal off any suspected leaks in the hull of the Spektr module. Also, four environmental effects payloads were recovered from the external hull of the space station. Over four tons of materials and experiments from the SPACEHAB Double Module in the Shuttle's payload bay were transferred to *Mir*, including nearly 2,000 pounds of water. Among the material items replaced were a new motion control computer, an old Elektron oxygen generator, batteries and three air pressurization units complete with air.

On 3 October 1997, the spacecraft *Atlantis* undocked from *Mir* and carried out a 46-minute manoeuvre around the space station. As they approached the suspected leaking area of the station, two of the Russian cosmonauts on *Mir* opened a pressure valve to see if the crew members on the shuttle could detect any seepages – but none were detected. A number of other minor experiments were carried out aboard *Atlantis*, including the continuing Radiation Monitoring Experiment, the Commercial Protein Crystal Growth investigation and the Shuttle Ionospheric Modification with Pulsed Local Exhaust. The Orbiter was scheduled to land on 5 October but because of low clouds over the Cape, the mission was extended by one day. On 6 October the spacecraft *Atlantis* touched down at the Kennedy Space Center. This was the last flight of *Atlantis* before being returned to Rockwell for maintenance and refurbishment.

Another cargo spacecraft reached *Mir* on 8 October, *Progress M-36*, which was launched from Baikonur on a Soyuz-U rocket with more supplies of food, water, oxygen and equipment. After docking in the aft port of the Kvant 1 module it was emptied and as usual, over the next couple of months the spacecraft was refilled with rubbish and any unwanted items. After undocking on 17 December it was placed in a decaying orbit to be burnt up on re-entry in the Earth's atmosphere on 19 December. This was followed one day later by *Progress M-37* on 20 December with more supplies and no doubt some Christmas surprises. The cargo spacecraft docked on 22 December in the aft port of the Kvant 1 module. After being unloaded it was refilled with the usual waste and unwanted articles.

The new year started with the arrival of the American Space Shuttle *STS - 89* (*Endeavour OV-105*) on 20 January 1998, to bring a new crew for the Russian space station. Among the crew was Salizhan Shakirovich Sharipov (Mission Specialist – Russian Space Agency). Part of the mission was to deliver the relief crew member for *Mir*, Andrew S.W. Thomas (Mission Specialist and Cosmonaut Researcher), and to return David A. Wolfe (Mission Specialist and Cosmonaut Researcher).

The launch of *STS-89* was first scheduled for 15 January 1998, but with the return of the Orbiter *Endeavour* after it had been serviced and refurbished by Rockwell, it was decided to use *Endeavour* instead of *Discovery*, so the launch was put back until 20 January. Then the Russian Space Agency requested a further delay because they wished to complete a series of experiments on board *Mir* before the two spacecraft docked. The launch was then re-scheduled for 22 January. At 21.48 hours EST the sky around Cape Canaveral was lit up when *STS-89* lifted off the launch pad and into space.

After two days in orbit and at an altitude of 214 miles above the Earth, *Endeavour* manoeuvred into position alongside *Mir* and the two spacecraft docked. The transfer of the two American astronauts was about to begin, when Andrew Thomas showed concern that his Russian Sokol pressure suit did not fit. The transfer was held up whilst adjustments were made to the suit and only after they had been made to the satisfaction of Thomas, was the transfer allowed to go ahead. The pressure suit was needed only if a situation arose and the crew of *Mir* had to return to Earth aboard a Soyuz spacecraft.

During the transfer of the two astronauts, over 8,000 pounds of water, logistical hardware and scientific experiments were also transferred during a four day period. A number of experiments and redundant hardware was also transferred from *Mir* to *Endeavour*.

The Orbiter *Endeavour* returned to Earth on 31 January 1998 after spending close to nine days in space and completing 139 orbits of the Earth.

On the 30 January 1998, *Progress M-37* was undocked and placed in a decaying orbit to make way for the arrival of *Soyuz TM-27* which had been launched on a Soyuz U2 rocket from the Baikonur Cosmodrome the day before. The crew of Talgat Musabayev, Nikolai Budarin and Leopold Eyharts (France), docked in the aft port of the Core module. They were to join fellow cosmonauts aboard *Mir* and it was initially hoped that *Soyuz TM-27* would be able to dock with *Mir* whilst the American space shuttle *Endeavour* was there. If all three crews had been aboard the space station at the same time there would have been thirteen astronauts/cosmonauts, setting a new on-board world record for space crews. But the French space agency, having paid a considerable sum of money to the Russian space agency, vetoed this, as in their opinion it could jeopardize their cosmonaut's Pegase experiment and that was more important. The experiment took a matter of two weeks to complete and on 19 February 1998 Leopold Eyharts, together with *Soyuz TM-26* cosmonauts Alexsandr Solovyov and Pavel Vinogradov, undocked from *Mir* and returned to Earth, landing near Arkalyk, Kazakhstan. The following day American astronaut Andy Thomas and the crew of *Soyuz TM-27* undocked from the Kvant module port and re-docked with the forward port of *Mir*. This left the Kvant module port free for the *Progress M-38* unmanned cargo spacecraft which was launched on a Soyuz V rocket and arrived on 14 March. This enabled all the supplies to be placed into the Kvant module where they were to be kept. It also carried the VDU-2 propulsion unit which was installed later. After the cargo spacecraft had been unloaded, it was refilled with waste and undocked on 15 May to be placed in a decay orbit to be burnt up on re-entry.

An EVA was carried out on 6 April to complete the repairs to the damaged Spektr solar panel. Once this was done they began a series of three EVAs during which the new VDU-2 station orientation engine was installed into the Sofora boom. The old engine was removed and left to drift to become part of the ever increasing 'scrapyard' in space.

The next unmanned cargo spacecraft, *Progress M-39*, was launched on 14 May 1998 and docked with *Mir* on 16 May in the aft docking port of the Kvant 1 Module. After being unloaded it was routinely refilled with waste over the next few months, then on 12 August, it was undocked to make way for the arrival of *Soyuz TM-28* on 13 August and placed in an orbit well away from the space station.

The launch of *STS-91* (*Discovery OV-103*) was the ninth docking mission between the Space Shuttle and *Mir* and also marked the return of the last American astronaut to work aboard the Russian space station. With the exception of a ten-minute delay, due to fuel loading in the Space Shuttle's external tank, the launch went ahead on schedule and *STS-91* lifted off the launch pad on 2 June 1998. After two days in orbit, *Discovery* edged her way alongside the Russian space station and docked at an altitude of 208 miles above the Earth. As the hatch was opened the first to greet the *Discovery*'s crew was American astronaut Andrew Thomas who had spent 130 days on the space station. Together with the other members of *Discovery*'s crew, was Russian Mission Specialist Valery Victorovitch Ryumin who entered the space station to greet his fellow Russian cosmonauts. Over the next four days more than 1,100 pounds of water and nearly 4,700 pounds of technical and scientific equipment were transferred between the two spacecraft. This included the long-term United States experiments that had been ongoing with every American crew member. These included Human Life Sciences and the Space Acceleration Measurement System as well as two crystal growing experiments. The two spacecraft separated on 8 June, bringing the Shuttle–*Mir* docking program to an end. This mission was also to be the last one to collect an

American astronaut from the Russian space station and the end, for the time being, of a very close co-operation between the two space-dedicated countries.

Back in free space the crew on *Discovery* continued to carry out further experiments, such as the Alpha Magnetic Spectrometer (AMS). This was the first mission of this delicate piece of equipment that was designed to look for dark and missing matter in the universe. Data received from AMS was to be transmitted back to Earth through the KU-band communications system aboard *Discovery*, but a fault appeared so the information had to be recorded aboard the spacecraft. A bypass system was then set up, that allowed the recorded data to be down-linked to Earth via S-band F/M communications as *Discovery* passed over a ground station. With all these experiments completed and having spent almost ten days in space and completing 155 orbits of the Earth, *Discovery* touched down at the Kennedy Space Center on 12 June 1998.

The future of the *Mir* space station was now under serious review. It had gone well past its allotted five-year program and was now approaching its tenth year in space. *Progress M-39*, which was still docked in the aft docking station of the Kvant 1 module, was undocked on 11 August so as to allow for the arrival on 13 August of *Soyuz TM-28*, launched on a Soyuz-U rocket on 11 August with cosmonauts Gennadi Padalka, Sergei Avdeyev and Yuri Baturin. With the new crew safely onboard, the normal housekeeping duties were carried out along with a number of astrophysical and geophysical experiments. Yuri Baturin was the first Russian politician to go into space and was to spend twelve days aboard the space station carrying out research.

Crew of *Soyuz TM-28* in their training capsule. L-R: Sergei Avdeyev, Gennady Padelks and Yuri Baturin. (Roscosmos)

On 25 August 1998, *Soyuz TM-27* crew Talgat Musabayev, Nikolai Budarin and Yuri Baturin, undocked from *Mir* and returned to Earth, landing in Kazakhstan.

The following month on 1 September 1998, the unmanned Progress spacecraft, *Progress M-39*, was re-docked with the aft docking station on the Kvant 1 module following the re-docking of *Soyuz TM-28* to the Core module. Amongst the fuel and supplies that were transferred was the Znamya-2 solar illumination experiment. This experiment, fixed to the nose of the cargo spacecraft, was first tried out in 1962 and designed to reflect sunlight onto selected cities in Russia. Unfortunately at the time its antenna did not deploy despite a number of attempts, so the experiment was abandoned. Again after a number of attempts the antenna once again failed to deploy, so the whole project was abandoned once again. After undocking from *Mir* on 25 October, the cargo spacecraft, after being filled with unwanted materials and rubbish, was placed into a decaying orbit to be burnt up on re-entry over the Pacific Ocean four days later.

The Spektr module of the space station was still giving some cause for concern, so on 15 September 1998 Gennadi Padalka and Sergei Avdeyev donned EVA space suits, and depressurized the module. After carrying out a detailed inspection and finding nothing wrong, they powered the module back up and then re-pressurized it. The whole of the space station was now up and running.

More supplies arrived with *Progress M-40* on 27 October on a Soyuz-U companion rocket and docked in the aft port of the Kvant 1 module. The cargo contained water, oxygen, fuel, personal items and extra software, along with the normal food supplies. Also inside the cargo spacecraft was the Znamya 2.5 experiment. This was another attempt using a solar mirror to beam sunlight back to Earth during the hours of darkness – despite the best of efforts it was unsuccessful.

One of the most significant moments in the history of international space co-operation happened on 20 November 1998. The first section of the ISS, the control module named Zarya (Sunrise), was launched from the Baikonur Cosmodrome in Kazakhsta. The 44,000lb module, also known as the Functional Cargo Block (FGB) by the Russians, was launched into a 210 nautical mile altitude by a 3-stage Proton rocket. Built by the Khrunichev State Research and Production Space Center (KhSC) in Moscow, who were under subcontract to the Boeing Company, the module was funded by NASA. The Zarya module was to provide electrical power, communications and orientation control when connected to the next module to be attached – Unity. Unity was taken into orbit by *STS-88* Space Shuttle Orbiter *Endeavour* (OV-105) on 4 December 1998, where the two modules were mated and stayed in orbit, awaiting the arrival of the Service Module which was launched from Kazakhstan. Zarya's two solar arrays, each of which were thirty-five feet long and eleven feet wide and had six nickel-cadmium batteries, provided three kilowatts of electrical power. On the arrival of the Service Module, the Zarya carried out an automated and remotely piloted docking with it and with its docking ports enabled, unmanned Progress cargo spacecraft and Soyuz spacecraft were able to dock. The Service Module took over the role of Zarya in supplying electrical power,

Soyuz TM-28 launching from Baikonur. (Roscosmos)

Zaraya, the first component of the ISS, being launched. (RKS Energia)

The Zarya module. (NASA)

communications and orientation control, allowing Zarya to be turned into a storage and external fuel tank module.

Russian cosmonauts were becoming more and more part of the Shuttle program and the inclusion of Sergei Krikalev showed that the co-operation between the two countries was working. The launch of *STS-88*, which had been originally scheduled for 3 December, had to be postponed for 24 hours when time ran out on the launch window. At T-4 minutes, when the Orbiter's hydraulic systems were switched on, a master alarm that was associated with No.1 hydraulic system was activated. The Shuttle's engineers were quickly into the problem but at T-31 seconds time ran out and the launch was aborted until 4 December 1998. At 03.35 hours EDT the skies around the Kennedy Space Center were lit up as the Space Shuttle's engines burst into life and seconds later lifted the spacecraft off the launch pad and into space.

The primary objective on the mission was to start the assembly of the International Space Station (ISS). This was made up of two modules: Zarya which was owned by NASA, but had been built in Russia, and Node-1 (Unity) which was built and owned by NASA. The two modules were connected by PMA-1 (Pressurized Mating Adaptor), a docking module. On the morning of 5 December *Endeavor*'s docking system was connected to the 12.8-ton Node-1 (Unity) module using PMA-2. The following day, the Remote Manipulator System arm was used to grasp the Zarya module from orbit, which was then mated to Unity (Node-1). There were some initial problems when it was discovered that the Unity/Zarya fittings would not align properly and had to be separated for adjustments to be made. Astronauts Jerry Ross and James Newman then commenced the first of three EVAs to attach power cables, connectors and handrails. With all the cables and connectors in place the modules were powered up. Problems again were encountered when some of the floodlights failed, plunging areas of the modules into complete blackness, but these were soon rectified.

The two astronauts during the second and third EVAs evaluated a Simplified Aid for EVA Rescue (SAFER) unit. This was a self-rescue device to be used in the event a space walker should become detached from the spacecraft during an EVA. They also prepared the Unity module for further future additions by removing restraint pins on the four hatchways. A sunshade was installed over the two data relay boxes to protect them from the direct sunlight and then they carried out a complete photographic survey of the space station.

On 10 December American astronaut Robert Cabana and Russian cosmonaut Sergei Krikalev carried out an IVA (Internal Vehicular Activity) and entered the new space station, followed closely by Jerry Ross and James Newman.

A number of secondary payloads were evaluated including the successful deployment of the Shuttle's KU-band antenna and the Hitchhiker payload, which involved the Mighty-Sat and SAC-A satellites. With all the experiments completed and the successful docking of the two modules, the Orbiter *Endeavour* prepared for re-entry. On 15 December 1998, after almost twelve days in space and 186 orbits of the Earth, the spacecraft *Endeavour* touched down at the Kennedy Space Center.

One of the last manned spacecraft to visit *Mir* blasted off the pad at Baikonur Cosmodrome, Kazakhstan on 20 February 1999. *Soyuz TM-29* with Russian cosmonaut Viktor Afanasyev, Slovakian cosmonaut Ivan Bella and French cosmonaut Jean-Pierre Haignere aboard, docked with the space station on 22 February. A number of new experiments were carried out, but the majority of the work consisted of finishing off existing experiments and investigations that had been left behind by previous crews, and preparing the space station for shut down. On 27 February, *Soyuz TM-28* commander Gennadi Padalka and the Slovak cosmonaut Ivan Bella undocked from *Mir* and returned to Earth, landing in Kazakhstan the following day.

The three remaining cosmonauts, Viktor Afanasyev, Sergei Avdeyev and Jean-Pierre Haignere, carried out a series of spacewalks examining the exterior of the space

station. On 2 April the cargo spacecraft *Progress M-41* was launched on a Soyuz-U rocket and two days later docked in the forward port of the Core module with fresh supplies and materials. After being unloaded, the crew refilled it over the next couple of months with all their unwanted rubbish and on 17 July it was undocked and placed into a decaying orbit to burn up on re-entry over the Pacific Ocean. The crew continued to carry out their experiments and research until the arrival of the next cargo spacecraft *Progress M-42* on 16 July. After docking in the aft port of the Kvant 1 module it was unloaded and refilled with waste and all the uncompleted experiments and investigations. It was undocked on 2 February 2000 and placed in a decaying orbit to burn up on re-entry. The space station was then powered down and then placed into a free drift mode. Then, on 27 August 1999, the hatch was closed on the Russian space station *Mir*. The three cosmonauts landed in Kazakhstan on 28 August 1999 and for the first time since September 1989 there were no human beings in space.

The launch of *STS-96* (*Discovery OV-103*) created a new milestone in the rapidly developing international space program and will go into the history books as being the first Space Shuttle to dock with the International Space Station (ISS). The launch had originally been scheduled for 20 May, but after a heavy hailstorm whilst the Shuttle was on the launch pad, damage was discovered to the foam insulation that covered the large external tank. The orbiter *Discovery* was then taken back into the Vehicle Assembly Building (VAB) for examination and some repairs to be carried out. A month earlier, on 20 April, a fault had appeared during interface testing, when it was discovered that an electrical cable between the left Solid Rocket Booster (SRB) and the electronic assembly boxes was not conveying the necessary signals, so the cable had to be replaced. *STS-96* was launched on 27 May 1999 and amongst the crew was Russian Mission Specialist Valery Ivanovich Tokarev.

After reaching orbit, *Discovery* rendezvoused with the ISS and carried out docking procedures with the PMA-1 docking unit. The following day astronauts Tamara Jernigan and Daniel Barry entered the tunnel adapter hatch into the payload bay. During their 7 hour and 55 minute EVA, they transferred equipment to the exterior of the ISS. This consisted of the Integrated Cargo Carrier (ICC) which carried the Russian cargo crane STRELA and was mounted on the exterior of Zarya, the Russian section of the station; the US built crane, the Orbital Replacement Unit (ORU) also known as the Orbital Transfer Device (OTD); and the SPACEHAB Oceaneering Space System Box (SHOSS).

On 31 May the hatch to the Unity module was opened and the crew began to transfer equipment from the Orbiter to the module. This consisted of communications equipment and battery units. Once completed the crew then fitted sound insulation to the Zarya module. It was to be another three days before the transfer of equipment was completed, and on 3 June, the Orbiter *Discovery* undocked from the ISS. With all the experiments completed the Orbiter *Discovery* headed back toward Earth. Having completed 153 orbits of the Earth and after almost ten days in space, the spacecraft touched down at the Kennedy Space Center on 6 June 1999. This was the 11th night landing of the Space Shuttle.

CHAPTER EIGHT

Into the Millennium

No less than 150 missions to the ISS, including unmanned cargo spaceflights, were scheduled between 2000 and July 2024, when additional laboratory and habitation modules were added, along with additional equipment necessary for the survival of the occupants.

Although missions to the ISS were now being planned, other missions into space had not been forgotten and the deployment of the Chandra spacecraft by *STS-93* was one of them.

Because of the slowly decaying orbit of the *Mir* space station, on 1 February 2000 the Russians launched another unmanned cargo spacecraft, *Progress M1-1* on a Soyuz U2 rocket, which docked in the aft Kvant 1 module and pushed the *Mir* space station into a higher orbit. The main reason for all this, was because the Russians had still not decided what to do with the space station. On the 3 April 2000, the privately funded spacecraft *Soyuz TM-30* blasted off the launch pad at Baikonur Cosmodrome with two cosmonauts, Sergei Zalyotin and Alexsandr Kaleri, aboard. After docking on 6 April in the aft port of the Core module, the two crew members entered the deserted space station and reactivated the life support systems. Their mission was to link up with *Mir* in an attempt to carry out repairs on a suspected air leak and carry out a condition assessment. The flight had been financed by MirCorp, a consortium of Western businessmen, to the tune of $20 million, who had thoughts of turning the now defunct space station into the world's first space hotel. The dream of creating a space hotel ended when funding

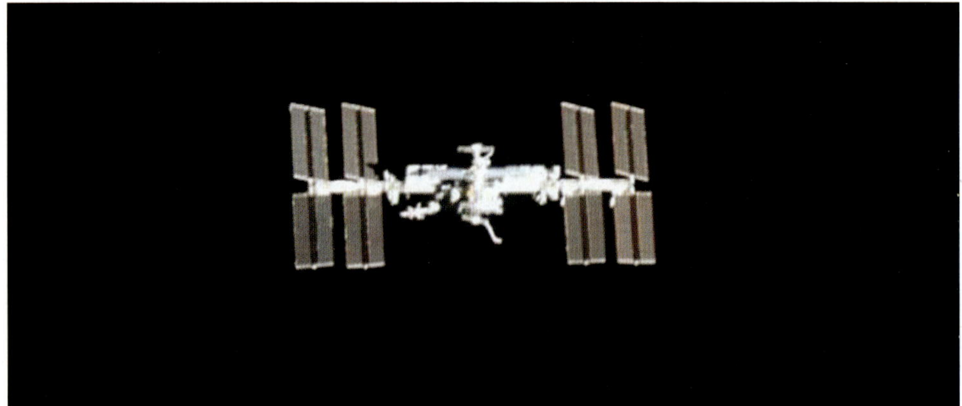

The International Space Station. (NASA)

ran out and ultimately was instrumental in its demise. On 25 April *Progress M1-1* was undocked from *Mir* to make way for *Progress M1-2* which was launched on 25 April from Baikonur on a Soyuz U2 rocket. *Progress M1-2* docked in the aft Kvant 1 port and contained food and supplies for the crew. *Progress M1-1*, after being undocked, was placed in a decaying orbit to be burnt up on re-entry. After a detailed examination of the now aging space station, that despite selling seats and space on it, the cost of maintaining *Mir* was deemed to be too expensive, so the project was abandoned.

Launched to the ISS on 19 May 2000, *STS-101* (*Atlantis OV-104*) went into orbit at an altitude of 173 nautical miles. The crew of James Halsell (Commander), Scott Horowitz (Pilot), Mission Specialists Jeffrey Williams, Mary Ellen Weber, Janice Voss and Susan Helms, included Russian Mission Specialist Yuri Usachev. *Atlantis* docked with the ISS in the forward port of PMA-2 (Unity) on 21 May. The mission had a number of objectives including the taking of air samples, monitoring carbon dioxide and replacing all the air filters within the space station. The crew then replaced four suspect nickel-cadmium batteries, replaced the memory unit in the radio telemetry system and replaced the port

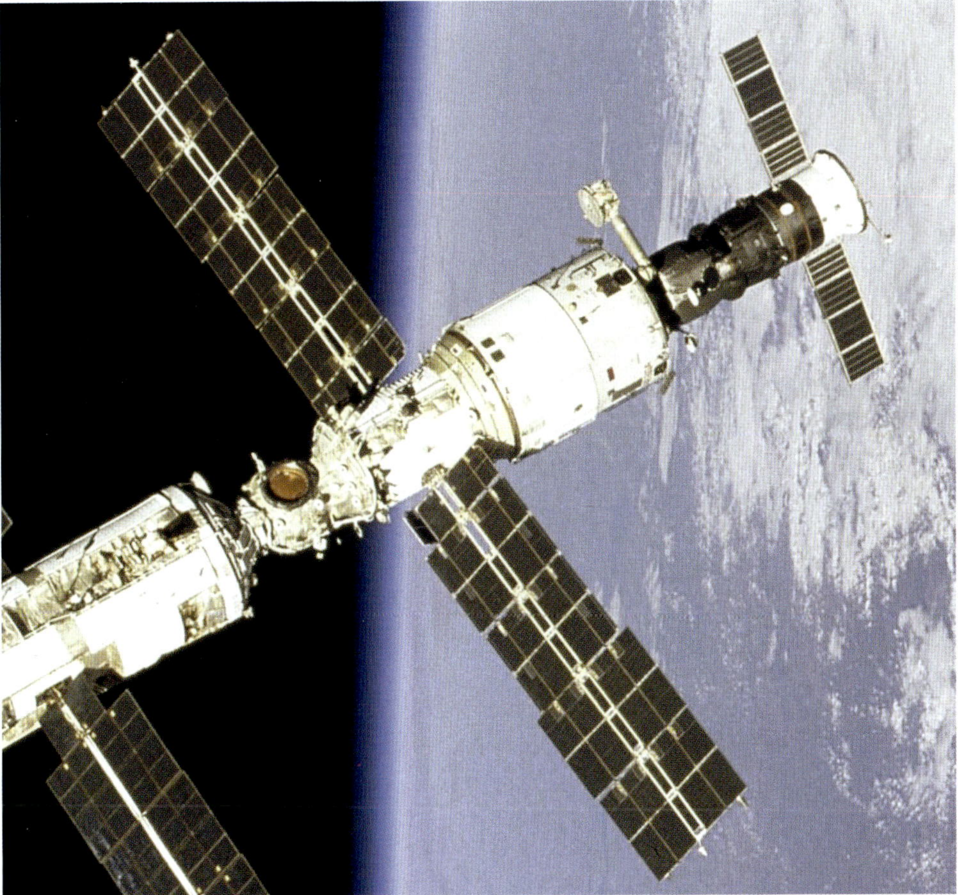

Zvezda module connected to the ISS. (NASA)

early communications antenna. They also took on board US and Russian hardware, provided EVA tools, and stowed the equipment and provisions for future missions. Amongst the supplies unloaded were water and water transfer equipment, film and videotape, a number of office supplies, personal equipment, medical supplies and exercise equipment. The latter was one of the most important items, as the Russians had discovered during their lengthy stays in their space station *Mir*. The problems cosmonauts/astronauts experienced on their return to Earth would have been greatly alleviated had they been given a strict program of exercise whilst in the weightless space environment.

The decision having been taken regarding *Mir*, the two Russian cosmonauts, Sergei Zalyotin and Alexsandr Kaleri, shut down the space station and closed the hatch for the last time on 15 June and returned to Earth on 16 June, landing near Arkaylk, Kazakhstan.

With *Mir* now empty and being prepared for deorbiting, attention was now concentrated on the International Space Station. On 6 August 2000 an unmanned cargo spacecraft, *Progress M1-3*, was launched and docked two days later in the aft port of the Zvezda module of the ISS. As the ISS did not have a resident crew, Progress

Yuri Malenchenko taking photographs from one of the cupolas on the ISS. (NASA)

Soyuz TM-31 being transported to the launch pad on a damp and foggy morning. (Roscosmos)

remained docked until the arrival of the American Space Shuttle *STS-106*, when it was unloaded by the crew. It was undocked remotely on 1 November and allowed to go into a decaying orbit to be burnt up on re-entry.

Launched on 8 September 2000 from the Kennedy Space Center, the Shuttle *STS-106* (*Atlantis OV-104*) was another of the Shuttle flights in the development of the ISS. It utilized the SPACEHAB double-module and the Integrated Cargo Carrier (ICC) and took supplies to the ISS in preparation for the October flight from Russia for the first residential crew. During their stay, the crew of Terrence W. Wilcutt (Commander), Scott D. Atlmann (Pilot), Mission Specialists Daniel C. Burbank, Edward T. Lu, Richard A. Mastracchio, Yuri Ivanovich Malenchenko (R) and Boris V. Morukov (R) unloaded supplies from both the Shuttle and the Russian *Progress M1-3* cargo spacecraft which had been coupled up to the ISS earlier in the year. The unloading of the supplies from the Russian cargo spacecraft was supervised by Dr Boris Morukov whilst the other Russian cosmonaut, Yuri Malenchenko, a Colonel in the Russian Air Force, later carried out only the second Russian/American spacewalk. The spacewalk was to hook up Communications, Telemetry and electrical cables between the Zvezda Module and the Zarya Control Module. One of the last jobs to be completed by this mission was the installation of the first toilet aboard the ISS. On 19 September, after almost 12 days in space, the Orbiter *Atlantis* touched down at the Kennedy Space Center after an almost perfect mission.

Progress M1-2 was undocked from the *Mir* space station on 15 October and placed into a decaying orbit, to make way for *Progress M-43* which was launched on a Soyuz-U rocket the following day. It docked in the Kvant module port four days later to boost the *Mir* space station into a higher orbit, but the declining fortunes of the Russian Federation (the former Soviet Union) had restricted future visits to *Mir* to one per year and that was deemed to be an unnecessary expense so *Mir* was left to deorbit in 2001. *Progress M-43* was fully loaded with supplies, but was never unloaded because there was no crew aboard. This then begs the question why it was ever sent. It was remotely undocked on 25 January 2001 and placed in a decaying orbit to enter the Earth's atmosphere on 29 January where it burnt up over the South Pacific Ocean.

The next mission into space was for the continuing in-orbit assembly of the International Space Station. *Soyuz TM-31* (*Expedition 1*) was launched from Baikonur on 31 October 2000 and docked two days later at the aft port of the Zvezda module on the ISS. The crew of Yuri Gidzenko, Sergei Krikalyov and William Shepherd then became the *Expedition 1* Crew and were to remain aboard the ISS for a number of months, carrying out numerous experiments and general maintenance of the space station. After transferring the crew and supplies, the spacecraft undocked and moved around the ISS to the other side, allowing the port to be free for the unmanned cargo spacecraft *Progress MI-4* to dock in the nadir port of the Zarya Module. Their spacecraft *Soyuz TM-31* remained connected to the space station as their rescue vehicle.

Progress MI-4 attempted to lock into the nadir port of the Zarya module, but the Kurs docking system failed, causing the manual TORU system to be used. Once unloaded and refilled with waste and other unwanted materials, the cargo spacecraft was undocked.

The *Expedition 1* crew to the ISS. L-R: Sergei Krikalev, William Shepherd and Yuri Gidzenko. (NASA)

The space station *Mir* breaking up as it passes through Earth's atmosphere. (NASA)

It spent the next twenty-five days in free flight before being re-docked in the Zarya module. It was finally undocked and placed into a decaying orbit on 8 February. In the meantime another cargo spacecraft, *Progress MI-5*, was launched on a Soyuz-U rocket on 24 January, docking in the Kvant module of *Mir*. Its only task was to lock on to the now defunct space station and put it into a decaying orbit. Onboard the ISS, the crew moved the rescue vehicle, *Soyuz TM-31*, from the Zvezda aft port to the nadir port of the Zarya module, a manoeuvre that took 1 hour and 31 minutes. This left the docking port open for the next cargo spacecraft, *Progress M-44*, which was due to arrive on 26 February 2001.

Launched on 8 March 2001, the American Space Shuttle, *STS-102* (*Discovery OV-103*), carried amongst the crew two American and one Russian mission specialists, James Voss, Susan Helms and Yuri Usachev (R), who were to replace the three resident astronaut/cosmonauts. Their primary objective was connect an Italian Logistics Module to the space station. After docking with the ISS, a number of experiments were carried out concerning the payloads aboard the spacecraft and repairs were carried out on a satellite. *Discovery* provided transportation back to Earth for the three *Expedition 1* crew members, William Shepherd, Yuri Gidzenko and Sergei Krikalev, who had arrived aboard *Soyuz TM-31*, and who had been carrying out experiments on the space station.

In the meantime the *Progress M-44* cargo carrying spacecraft had been launched on a Soyuz-U rocket and docked in the Zvezda module of the ISS, delivering supplies, experiments and mail to the ISS crew. As with almost all the cargo spacecraft, once they had been unloaded they were refilled with unwanted materials and rubbish and placed in a decaying orbit to be burnt up on re-entry.

After much speculation as to when the now empty *Mir* space station was to be allowed to deorbit and be destroyed during re-entry over the South Pacific, it was finally placed in a controlled decaying orbit on 23 March 2001. In a lighter moment the fast food chain Taco Bell promised every American a free taco – if the core of *Mir* hit a floating target in the South Pacific. The target, a 144 square metre vinyl sheet

The redundant space station *Mir* breaking up as it enters Earth's atmosphere.

with a bulls eye and the words 'Free Taco here' in bold purple letters, was anchored about 15km off the Australian coast – thousands of kilometres to the west of the expected landing area. The cost to Taco Bell, if the core part of *Mir* did hit the target, was estimated to be around $10 million. Fortunately for them it missed by a huge margin. William Ailor, who was director of the Aerospace Corp's Center for Orbital Re-entry Debris Studies at the time, was quoted as saying that the odds of *Mir* hitting the target were slim to none – he was right.

Russian cosmonauts continued to fly aboard the American space shuttle, and on 19 April 2001, *STS-100* (*Endeavour OV-105*) lifted off the launch pad at the Kennedy Space Center to rendezvous with the ISS. Amongst the crew of Kent Rominger (Commander), Jeffrey Ashby (Pilot), Mission Specialists Chris Hadfield (Canada), John Philips, Scott Parazynski and Umberto Guidoni (Italy), was Russian Mission Specialist Yuri Lonchakov. The main function of the *STS-100* mission was to service the ISS and deliver three robotic components to be installed during two spacewalks. The Canadian Robotic Arm – Canadarm 2, was fitted to the platform that had been installed by the previous crew. The crew also delivered supplies and brought back completed scientific experiments

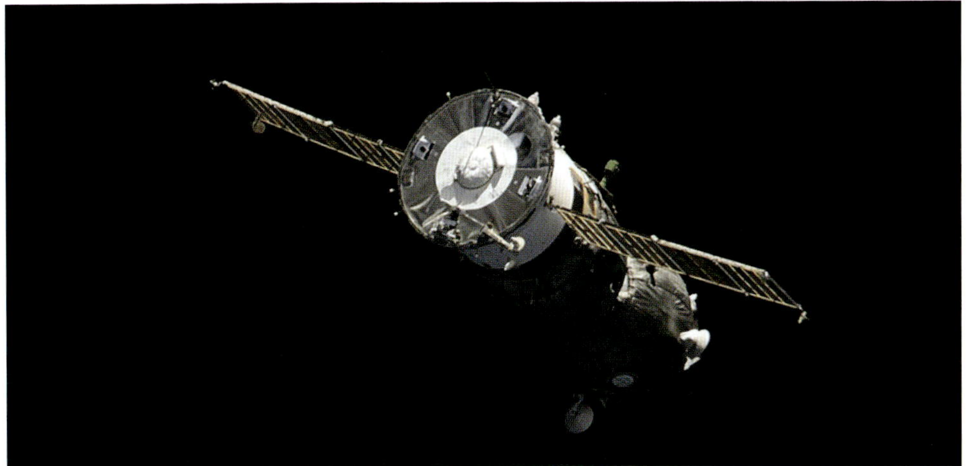

Soyuz TM-32 approaching the ISS. (Roscosmos)

and waste products. The crews were now becoming more and more international, with an Italian and Russian on board as Mission Specialists. The Rafaello logistics module was attached to the US laboratory Destiny, then four days later returned to the cargo bay of *Endeavour*. The Russians, despite their now being heavily involved with NASA, still launched their own spacecraft to the ISS.

On the 28 April 2001 *Soyuz TM-32* (*Expedition 2*) lifted off the pad at Baikonur and docked two days later in the nadir port of the Zarya module. The American Space Shuttle *STS-100* had undocked from the space station a few hours earlier to allow for the arrival of the Russian Soyuz spacecraft. *TM-32* was to provide a fresh lifeboat for the International Space Station as the existing lifeboat was deemed to be at the end of its in-space storage. This flight also caused a great deal of controversy because, besides the cosmonauts Talgat Musabayev and Yuri Baturin, the Russians took a paying American tourist 'Dennis Tito' aboard their spacecraft. The Americans objected on the grounds that the ISS was a scientific space station not a tourist hotel. It has to be said that the millions of dollars Tito paid for the flight helped the Russian Space Agency's sadly depleted funds. One week later the crew of *TM-32* transferred their contoured seats to Soyuz *TM-31* and returned to Earth in it,leaving *TM-32* as the ISS rescue vehicle.

The next to visit the ISS was *STS-105* (*Discovery OV-103*), on 12 August 2001. Launched on 10 August, the orbiter docked in the forward port on the Destiny 2 module. The primary objective of the mission was to transport the *Expedition 3* crew of Mission Specialists Frank Culbertson (A), Vladimir Dezhurov (R) and Mikhail Tyurin (R) to the ISS, together with a number of scientific instruments and experiments. Supplies for the crews on the space station were also taken and waste was removed. The *Expedition 2* Crew of Yuri Usachev, Susan Helms and James Voss, who had been on the space station for five months, returned with *Discovery*.

More supplies were delivered to the ISS on 23 August, when the cargo spacecraft *Progress M-45* docked in the Zvezda module. It was unloaded and refilled with the usual

Space tourist Dennis Tito enjoying an apple on his return to Earth. (Roscosmos)

waste, but remained docked until 21 November, when it was undocked to make way for the arrival of *Progress M1-7*. The *M-45* was then placed in a decaying orbit around the Earth where it burnt up on re-entry. A new module, Pirs, was delivered to the ISS on 17 September 2001, when *Progress DC-1*, launched on a Soyuz-U rocket, docked in the nadir port of the Zvezda module. The pressurized cargo module had been removed to accommodate the new Pirs module. It was undocked on 26 September and placed in a decaying orbit to burn up on re-entry over the South Pacific Ocean.

A few weeks later, on 21 October 2001, *Soyuz TM-33* (*Expedition 3*) was launched from Baikonur on a Soyuz-U rocket, docking in the Zarya module. The spacecraft carried two Russian Mission Specialists, Viktor Afanasyev and Konstantin Kozeyev, and one French cosmonaut, Claude Andre-Dehays, to the International Space Station (ISS). Their Soyuz spacecraft was to be left on the station as a lifeboat. The *TM-33* crew returned in *Soyuz TM-32* after spending eight days on the ISS. The *Soyuz TM-33* spacecraft was to remain attached to the space station as a lifeboat for the long-term crew that was remaining on the ISS.

On 26 November 2001, *Progress M1-7* was launched on a Soyuz-FG companion rocket from Baikonur, with much needed supplies and experiments. However, hard docking in the aft Zvezda module was delayed because of debris left behind in the docking port by the Kolibri-2000 when it had been deployed by *Progress M-45*. It wasn't until the 3 December that the cargo spacecraft finally docked, after the debris had been cleared during an EVA by the resident crew.

Launched on 5 December 2001, the *STS-108*, (*Endeavour OV-105*), was on an eleven-day mission to deliver hardware and supplies to the ISS and take a fresh crew, Mission Specialists Yuri Onufrienko (R), Daniel Bursch (A) and Carl Walz (A) to the space station and bring the returning crew of Frank Culbertson (A), Vladimir Dezhurov (R) and Mikhail Tyurin (R) back to Earth. Shuttles were becoming more and more like delivery trucks and buses. Once again the Rafaello logistics module was attached to the US laboratory Destiny and four days later returned to the cargo bay of *Endeavour*.

The launch of another cargo craft, *Progress M1-8*, on 21 March 2002, brought more supplies and materials to the resident crew of the ISS. After docking in the aft port of the Zvezda module, Progress was unloaded,and following the usual pattern, filled with waste. Another 'space tourist' visited the ISS on 25 April 2002 aboard the Russian spacecraft *Soyuz TM-34* (*Expedition 4*), launched on a Soyuz-U rocket, when South African Internet millionaire Mark Shuttleworth paid $20 million for the trip. He was placed aboard the spacecraft, together with Italian cosmonaut Roberto Vettori and Russian cosmonaut Yuri Gidzenko, as a mission specialist. Unlike the Dennis Tito trip, the inclusion of Mark Shuttleworth into the ISS crew was given approval by NASA because they now realized that it was one way for the Russians to boost their funding of the ISS. Their only reservations were about bringing untrained personnel into space, but Shuttleworth spent eight months training at Star City with other cosmonauts. During his time aboard he carried out a small research program studying ocean life and a number of biological experiments designed to help combat AIDS and other diseases. The three *Soyuz TM-34* cosmonauts returned to Earth on 5 May in the *Soyuz TM-33* spacecraft after spending ten days docked with the ISS, landing outside Arkalyk, Kazakhstan.

Then in May 2002 the Buran took centre stage once again, only this time not for the better. The last remaining full-scale test model was housed in a hangar at the Baikonur Cosmodrome, Kazakhstan, when the building collapsed. Eight people were in the building when it collapsed and were killed, the cause being put down to years of neglect and sub-standard workmanship. Then, with the sudden switch of power to Mikhail Gorbachov and his policy of Glasnost, there developed an openness that gave everyone a voice, and amongst these were those in the aerospace community. Space engineer and chief designer of the Vostok spacecraft, Konstantin Feoktistov, sent a nineteen-page document to Mikhail Gorbachov, criticizing the whole Buran program. He was supported by a number of other space scientists and engineers, and this in effect killed the program.

The missions to the ISS continued with the launch of the *STS-111* (*Endeavour OV-105*) mission on 5 June 2002 to deliver the new ISS crew of Mission Specialists, Valeri Korzun (R), Sergei Treschev (R) and Peggy Whitson (A), with a priority to replace the wrist roll joint on the Canadian Robotic Arm – Canadarm 2. The remote manipulator arm on *Endeavour* was used to remove the logistics module Leonardo from the spacecraft cargo bay and attach it to the Common Berthing Mechanism (CBM) on the underside of the US space laboratory – Destiny. Chang-Diaz and Perrin carried out a spacewalk to check the thermal coverings, they then carried out a second spacewalk to attach video, data and power cables to the Mobile Base Unit. They also re-located one of the external TV cameras. The wrist-roll joint on the Canadarm 2 was replaced during the third spacewalk.

The Leonardo logistics module was removed from the cargo bay of *Endeavour* and attached to the US laboratory Destiny. Four days later it was returned to *Endeavour*'s cargo bay. The returning ISS crew of Yuri Onufriyenko (R), Carl Walz (A) and Daniel Bursch (A) became part of the *STS-111* crew for the return to Earth on 19 June.

It wasn't until 25 June 2002 that *Progress M1-8* was finally undocked and placed in a decaying orbit to burn up on re-entry over the Pacific Ocean. The following day more supplies were delivered to the space station, when *Progress M-46* on a Soyuz-U rocket was launched. The cargo spacecraft docked in the recently vacated Zvezda module and, after the normal practice of unloading and refilling had been carried out, it was finally undocked on 24 September, but stayed in free flight until 14 October before entering the Earth's atmosphere and being burnt up on re-entry over the Pacific Ocean.

On the 10 November 2002, *STS-113* (*Endeavour OV-105*) lifted off the pad at Kennedy to carry a fresh crew to the ISS space station. The ongoing ISS *Expedition 6* crew consisted of Mission Specialists Ken Bowersox (A), Donald Petit (A) and Nikolai Budarin (R). On docking with the space station, supplies were transferred and over the next few weeks a number of experiments carried out, including the fitting of an integral truss and spacewalk work platform on the port side of the ISS. This would support an additional cooling system which would be needed when the solar array panels were extended to almost an acre. To enable the ISS crews to work more easily outside the space station, human-powered rail-cars called Crew and Equipment Translation Aid (CETA) carts would be installed to allow the EVA crews to traverse along the railway on top of the truss. The returning crew of Valery Korzun (R), Segei Treschev (R) and Peggy Whitson (A) embarked aboard *Endeavour* for the return to Earth after having been on the ISS since 10 June. The eleven-day mission was a complete success.

The next crew to the ISS were scheduled to fly on *Soyuz TMA-1*, which was a modified, updated version of the TM-34 Soyuz spacecraft. The original crew of Russian Sergei Zalyotin, Frank de Winne of Belgium and space tourist American pop star Lance Bass of the pop band 'N'Sync', had to be changed when a problem arose when the American was asked to leave Star City. This was because of his lack of dedication to the training that was vital to the mission, and the fact he was unable to come up with the $13 million required. Because of this and his 'cavalier' attitude toward the training, the Russians decided that he was too big a risk and dropped him from the mission to the ISS. Even when the crew went to the United States for familiarization training on the American segment of the ISS, they refused to take Bass, saying that he could only go if he paid his own way and got permission from NASA. Suffice to say he pulled out of the project. He did however get a 'Cosmonaut's Certificate' for the training he had completed. He had been so confident in going, that he had even had his name put on the crew patch. His place was taken by Yuri Lonchakov and the launch took place on 30 October 2002 from Baikonur Cosmodrome in thick fog. The spacecraft docked with the ISS on the nadir port of the Pirs module on 1 November. Frank de Winne and the other members of the crew carried out a program of twenty-three experiments covering life and physical sciences for the ESA (European Space Agency) and Roscosmos. They returned to Earth on 4 May 2003. Because of a technical malfunction to the Soyuz control system, the gentler controlled re-entry and descent reverted back to the harsher ballistic re-entry and descent. This resulted in a steep and off-target landing of the spacecraft. The craft landed 300 miles short of the planned area, and the crew was subjected to severe G-force. Communication with the Soyuz was lost because one antenna was ripped off during descent, and two more did not deploy. The crew regained communications through an emergency transmitter after landing. Due to this event, future crews were provided with a satellite phone to establish contact with recovery forces.

Lance Bass was not the only 'space tourist' to apply at the time. Polish businessman Leszek Czarnecki had also applied, but when his plans had been leaked to the press, plus the rising financial costs, he pulled out. In September the Chilean government entered into talks with the Russians with regards to sending one of their air force pilots into space. Their cosmonaut-candidate, Klaus von Storch Krüger, arrived in Moscow for a series of medicals to determine whether or not he was fit enough for a space flight. After successfully completing all the medicals, he was assigned for training for

Crew of *Soyuz TMA-1*. L-R: Yuri Lonchakov, Frank de Winne and Sergei Zalyotin. (Roscosmos)

Soyuz TMA-1 approaching the ISS. (NASA)

a flight aboard the *Soyuz TMA-2* (*Expedition 7*) spacecraft. Unfortunately the money ran out, and despite the Russian government's help, the Chilean government could not raise or justify the millions required and so the dream ended.

Plans were immediately put into action to replace the Chilean candidate with an experienced, trained Russian cosmonaut, Alexander Kaleri, but then all these plans were turned upside down with the devastating events of the 1 February 2003 – the loss of *STS-107* (*Columbia OV-102*) which disintegrated whilst on re-entry, killing all the crew. With the tragic news of the accident having stunned the world of space exploration, the *Expedition 6* crew completed their experiments and eight days later on 3 May, Kenneth Bowersox, Nikolai Budarin and Don Pettit undocked and returned to Earth aboard the *Soyuz TMA-1* spacecraft on the 4 May 2003, leaving *Soyuz TM-34* as the 'lifeboat' for the resident crew still aboard the ISS. These TMA flights were regarded as 'Taxi' flights to deliver supplies, crews and a fresh spacecraft.

On landing back on Earth the crew found themselves 460 kilometres off course, and search and rescue aircraft were sent to look for them. Two hours later they were spotted by one of the search aircraft, but it was to be a further three hours before rescue helicopters reached them. Initially the blame was placed on the cosmonauts for the spacecraft's deviation from the proposed landing site, or as one of the engineers put it, 'they pressed the wrong button', but later it was discovered that a fault in the spacecraft's computerized landing system was responsible. Later crews were issued with mobile phones enabling them to maintain contact with the recovery crews.

Soyuz TMA-6 approaching the ISS. (NASA)

It was now left to the Russians to keep the ISS running and on 2 February 2003,within hours of learning of the tragedy, they had launched *Progress M-47* on a Soyuz-U carrier rocket, to deliver one ton of food, equipment and mail to the crew still aboard the International Space Station. After docking in the aft port of the Zvezda module, the crew unloaded the cargo spacecraft and started to refill it with unwanted materials and waste. The tragic accident meant that the Russian cargo spacecraft would have to make more trips to the ISS than had originally been planned. This in itself caused a number of problems, because if Shuttle flights were suspended, only two Russian Soyuz TMA spacecraft, capable of carrying cosmonauts/astronauts, were being built a year. The cost of building these spacecraft was extremely high, approximately $18 million each, and the Russian Space Agency was already struggling to meet the cost of just these two, so to build more in a year was going to be extremely difficult unless additional funding was found. In addition to these financial concerns, was the cost of the disposable Progress cargo spacecraft that were only used just the once.

On 26 April 2003 *Soyuz TMA-2* was launched to relieve the crew aboard the International Space Station. Two days later the spacecraft docked in the nadir port of the Zarya module and after the usual safety checks the crew entered the space station. The incoming *Expedition 7* crew of Yuri Malenchenko (R) and Ed Lu (A) had originally been three, but because of the recent tragic event involving the American Space Shuttle, the third member of the crew, Alexander Kaleri, had be 'bumped'. All flights to the space station were having to be launched from Russia because of the 'hold'

Soyuz TMA-2 docking with the ISS. (NASA)

Crew of *Soyuz TMA-2*, R-L Yuri Malemchenko and Edward Lu. (NASA)

on all shuttle flights in the United States after the tragic accident. Both members of the *Expedition 7* crew were experienced cosmonaut/astronauts, Yuri Malenchenko had spent four months aboard *Mir* in 1994, and was aboard *STS-106* when it docked with the ISS for the first time. Ed Lu was also on that flight, so the two were not strangers, which helped considerably when it was realised that between the two of them they were going to have to continue to operate the science payloads already on board and maintain the space station. After a week of handover operations, the returning crew, Sergei Zalyotin, Frank de Winne and Yuri Lonchakov, boarded the *Soyuz TM-34* spacecraft and headed back to Earth. During the undocking procedure of *TM-34*, the computers aboard *Soyuz TMA-2* erroneously activated the spacecraft's thrusters causing the ISS's attitude control to move the space station 25 degrees off its planned orbit. A number of unplanned manoeuvres had to be carried out to put the space station back into its original position. The return to Earth was uneventful and the spacecraft touched down on 4 May near Arkalyk, Kazakhstan. The ISS program was now back on track.

On 10 August 2003, Yuri Malenchenko had set a space first when he married his fiance Ekaterina Dmitrieva. He was orbiting the Earth at the time and was over New Zealand, whilst his wife-to-be was in Texas, USA. Under Texas law it was permitted to carry out marriage ceremonies remotely. The ceremony went without a hitch.

Back on the ISS, the Progress cargo spacecraft *M-47*, having been refilled with waste, was undocked on 27 August and placed in a decaying orbit to be burnt up the following day as it passed through the Earth's atmosphere over the Pacific Ocean.

Soyuz TMA-3 about to dock with the ISS. (NASA)

Soyuz TMA-2 after landing with crew being recovered. (Roscosmos)

The next Russian mission to the ISS took place on 18 October just two days after the Chinese Taikonaut Yang Liwei returned to Earth after completing his country's first space flight. This mission was made under the most secret of missions and the results were not made public until after he returned in case there were problems.

TMA-3 was launched on 18 October 2003 on a Soyuz-FG carrier rocket, docking in the nadir port of the Pirs module. ESA astronaut Pedro Duque of Spain was accompanied by Michael Foale and Alexander Kaleri, the latter who, after being 'bumped' from the previous flight, was back on flight status. They carried out the normal housekeeping duties together with experiments, whilst Pedro Duque carried out his eight day visit conducting experiments connected with the ESA's Cervantes Mission. He returned with the *Expedition* 7 crew of Ed Lu and Yuri Malenchenko aboard *Soyuz TMA-2*. Michael Foale was the first American astronaut to serve on both the *Mir* and ISS space stations.

The Soyuz flights to the ISS were fast becoming a cause for concern amongst both RKK Energia and NASA. The whole space program in Russia was being run on a shoestring budget and money was fast running out. NASA wanted to divert some of its funding to the Russians in an effort to ease their situation, but the US Congress had prohibited the transfer of money to Russia because of their involvement in helping Iran develop its nuclear capacity. Just as in Russia, politics took centre stage and controlled the purse strings.

Russian space officials were becoming increasingly concerned regarding the future of the ISS, as it was becoming more and more difficult to find the money to send crews to maintain the space station. It was because of this problem that increasing interest was being shown by the Russians in taking paying 'space tourists' to the ISS, despite reservations being voiced by NASA. NASA's main concern on the other hand, was that because of the desperate financial situation, corners were being cut in the construction of the Soyuz spacecraft to the detriment of safety. But despite the financial worries the Russians insisted that corners were not being cut and that safety was their number one priority and continued to send supplies to the resident crews on the ISS.

On 8 June 2003, *Progress M1-10* was launched on a Soyuz-U carrier rocket from Baikonur with fresh supplies. The cargo spacecraft docked on 11 June in the nadir port on the Pirs module, where it was unloaded. Over the next few weeks it was filled with waste and unwanted materials and undocked on 4 September. Once undocked the spacecraft went into a slow decaying orbit before entering the Earth's atmosphere one month later on 3 October and burning up over the South Pacific Ocean.

Another cargo spacecraft, *Progress M-48*, was launched on 29 August 2003 on a Soyuz-U companion rocket and docked on 31 August in the aft port of the Zvezda module. After discharging its cargo, the cargo spacecraft stayed locked on until 28 January 2004, during which time it was refilled with unwanted materials and waste before it was undocked and placed into a decaying orbit where it burnt up on re-entry over the South Pacific Ocean. It was released in time for the arrival of the next cargo spacecraft, Progress M1-11, which was launched on 29 January 2004 on a Soyuz-U carrier rocket and, after a two-day flight, docked in the aft port of the Zvezda module. After being unloaded it was left docked until 24 May before being released to allow the next cargo spacecraft, *Progress M-49*, to arrive three days later on 27 May. Before that, *Soyuz TMA-4* (*Expedition 9*) blasted off on a Soyuz-FG rocket on 18 April 2004 from the same launch pad, 1/5 at Baikonur, that saw the launch of the first man in space – Yuri Gagarin – some forty-three years previously. Two days later the Soyuz spacecraft docked with the ISS in the nadir port of the Zarya module and delivered the relief crew of Mike Fincke (A) Gennady Padalka (R) and Andre Kuipers (Ndl). For Michael Fincke this was a milestone in his career when he became the first American astronaut to make his first spaceflight aboard a Russian spacecraft. The returning crew of Michael Foale (A) and Alexander Kaleri (R) also had Andre Kuipers (Ndl) who had spent only ten days aboard the space station, during which time he had performed a number of physiological and medical experiments for the ESA.

The continuous manning of the ISS was one of the most important factors in the space program. This was highlighted by Gennady Padalka, who when asked about the role of the astronauts on board the space station during an interview, said, 'We discovered an air leak in one of the outposts, which, had there not being a crew aboard to repair it, we could have lost the space station.' Whilst this was going on, the returning crew landed in Kazakhstan on 29 April.

Most of the known experiments carried out by the crews covered life sciences, the reactions of humans living in space and the problems associated with long periods of living in a weightless environment. Problems with fluid mechanics were slowly being

Soyuz TMA-4 on the launch pad at Baikonur about to be launched. (Roscosmos)

addressed by means of ingenious devices that sent back information to the scientists on the ground.

During their mission, the two-crew members carried out a spacewalk at the same time, leaving the space station unattended for the first time. Procedures covering the spacewalks had been worked out previously between the American and Russian controllers on the ground. Two spacewalks had been planned, but a third had to be carried out when an external gyroscope component on the US section went faulty. A problem then arose with the American spacesuit, making it unusable. The ground controllers and the two crew members got together and worked out a method whereby Michael Fincke could use the Russian Orlan-M spacesuit. It was successful, highlighting another area in which co-operation between the two countries was seen to be working.

After adjustments had been carried out on the Russian suit, Michael Fincke got to work and repaired the faulty spacesuit. Usually in a case like this, the faulty spacesuit would be exchanged for a new one when the next shuttle arrived. But because of the suspension of the shuttle program, the crew were having to rely on unmanned Progress cargo spacecraft for supplies and there was no room for a new spacesuit on board.

The next cargo spacecraft, *Progress M-49*, was launched on 25 May 2004 on a Soyuz-U rocket and docked in the aft port of the Zvezda module two days later with the usual supplies of food and other materials. After getting filled with the usual unwanted materials and rubbish, it was undocked on the 30 July and placed in a decaying orbit. Just under two weeks later another cargo spacecraft was launched on 11 August, *Progress M-50*, which docked in the aft port of the Zvezda module. It was unloaded and then over the next few months, like all the other cargo spacecraft, was filled with waste and unwanted materials. On 22 December it was placed in a decaying orbit around the Earth and left to burn up on re-entry. In the meantime there was another change of crew on 14 October, when *Soyuz TMA-5* (*Expedition 10*) was launched from Baikonur on a Soyuz-FG rocket. The routine launch suddenly turned into a minor drama, when the autopilot that was to guide the spacecraft through the approach and docking procedure malfunctioned as it approached the ISS. Salizhan Sharipov immediately took over manual control and guided the Soyuz spacecraft into the Pirs module docking port, confirming that the hours spent in the Soyuz simulator practicing manual control of the spacecraft had not been wasted! The relief crew of Leroy Chiao (A), Salizhan Sharipo (R) and Yuri Shargin (R) carried out the normal safety procedures and then, three hours after docking and all the routine tests had been carried out, the onboard crew 'cracked' the hatch and the two crews joined up.

Once on board the inbound crew spent the next eight days going through the safety procedures and housekeeping duties required to keep the ISS operational. Yuri Shagrin, the third member of the new crew, who was on his first spaceflight, carried out a number of experiments before embarking with the outgoing crew, leaving the remaining two to run the space station for the next six months. Repairs also had to be made to the Russian oxygen generator using spare parts brought by the ongoing crew. The problem with the cooling system in one of the American EVA suits also had to be sorted out. It wasn't until the *Expedition 9* crew of Mike Fincke (A), Gennady Padalka (R) and Yuri Shargin (R) had left on the Soyuz *TMA-4* spacecraft, which had been connected to the Zarya module, that the ongoing crew realised that the appetites of the returning crew had diminished the food supplies somewhat, and that ground control had not been informed of the problem. The *Expedition 10* crew had only brought a limited supply of additional food, so they had to ration themselves until the Russian *Progress M-51* unmanned cargo spacecraft was launched on 23 December 2004 on a Soyuz-U rocket, docking with the ISS in the aft port of the Zvezda module two days later. On board was 1,234 pounds of propellant, 110 pounds of oxygen, 926 pounds of water and 2,777 pounds of dry goods, which included 70 food containers. Had the Progress spacecraft been unable to dock, then Leroy Chiao and Salizhan Sharipov would have had to return to Earth at the beginning of January.

The *Expedition 10* crew were two of the most experienced to date. Leroy Chiao had been on three shuttle flights and spent over thirteen hours on two spacewalks whilst constructing the ISS. Salizhan Sharipov had spent more than 200 hours in space and was a mission specialist aboard *STS-89* on a flight to the space station *Mir*. During their stay aboard the ISS they carried out two spacewalks. These consisted of examining the exterior of the ISS and looking for any damage or deterioration of the structure. Leroy Chiao even managed to place his vote in the US Presidential elections by submitting an electronic ballot paper, by way of an electronic e-mail, to his county clerk's office – the first Presidential vote to be cast from space a mere 230 miles above the Earth.

Soyuz TMA-5 launching from Baikonur. (Roscosmos)

On 29 November the crew left the ISS and moved their *Soyuz TMA-5* spacecraft from the nadir port on the Pirs module to the Zarya module docking port. Although only 45 feet away, the manoeuvre was a very delicate operation, and consisted of backing the spacecraft away from the ISS for about 100 feet, then moving the spacecraft sideways for 45 feet, before carrying out the docking manoeuvre into the Zarya module. This left the Pirs docking module port free for the next ISS crew.

The ease at which the Russians and Americans now worked together symbolised the trust the two space agencies had in each other, something that would have been unheard of 20 years earlier. Since the tragic *STS-107* shuttle disaster, the Russian space agency had to be relied upon to carry the whole ISS program by themselves, which spoke volumes for the reliability of their rockets and spacecraft. But it wasn't until 2005 before the Shuttle program was up and running again.

The essence of these missions was to pave the way for future missions to distant planets such as Mars, and who knows maybe the outer planets and beyond. These missions of course will not take six months, more like six years, and the crews will have to be physically and psychologically ready for such missions. The information being collated from the experiments being carried out on the ISS was instrumental in ensuring that the crews would be fully equipped both physically and mentally to deal with long duration missions.

Progress M-52 was the next cargo spacecraft to be launched from Baikonur and it lifted off the launch pad on 28 February 2005 on a Soyuz-U companion rocket. It

Soyuz TMN-6 crew: L-R. Roberto Vittori (It), Sergei Krikalyov (R) and John Phillips (A). (NASA)

docked in the aft port of the Zvezda module on 2 March where it was unloaded. The cargo spacecraft remained there until 2 June, during which time it was refilled with waste and unwanted materials. It was then undocked from the space station and placed into a decaying orbit the following day and was burnt up as it entered the Earth's atmosphere over the Pacific Ocean.

On 15 April 2005, the *Expedition 11* crew of Sergei Krikalev (R), John L. Phillips (A) and Roberto Vittorio (ESA) was launched from Baikonur aboard *Soyuz TMA-6*, to relieve the resident crew of Leroy Chiao and Salizan Sharipov. Launched on a Soyuz-FG rocket, the spacecraft reached the space station after a two-day flight and docked in the nadir port of the Pirs module. The relief crew, after completing the obligatory safety checks, went aboard the space station on 17 April. On 24 April, together with the *Expedition 10* crew of Leroy Chiao (A) and Salizan Sharipov (R), Robert Vittorio returned to Earth aboard the Russian spacecraft *Soyuz TMA-5* after completing a number of experiments for the ESA. After a perfect re-entry and landing in Kazakhstan, because of the saturated landing area caused by melting snow and heavy rain the crew had to be airlifted by helicopter to Arkalyk to meet up with the other members of the recovery crew. This left just two crew members on the space station. There had been a few worrying moments for Leroy Chiao and Salizan Sharipov during their six and a half-month stay, especially when one of the main gyros that kept the ISS in a stable orbit, malfunctioned. Fortunately the two remaining gyros managed to keep the space station steady whilst the third gyro was repaired. The majority of the tasks carried out by the crew were housekeeping duties and the constant maintenance needed to keep the space station up and running. The crew also carried out a number of experiments, evaluations and tests connected to the effects of living and working in weightless conditions, including cardiovascular and physiology tests and observations.

Progress M-52 had been filled with trash, waste and items no longer needed, and released on 15 June to burn up in the Earth's atmosphere on re-entry over the Pacific Ocean. On 17 June 2005, the unmanned cargo spacecraft *Progress M-53* lifted off the pad at Baikonur on a Soyuz-U rocket, with supplies for the ISS. Everything went well until the final docking commands were needed and then the ground controllers found the spacecraft not responding. Sergei Krikalev immediately took over remote control using the TORU rendezvous system and eased the spacecraft into the aft docking port in the Russian Zvezda service module. The following day, Sergei Krikalev and John Phillips opened the hatch of the cargo spacecraft and removed the supplies. The supplies carried aboard the spacecraft amounted to 4,662lbs, of which 3,100lbs were dry cargo such as food, personal items, water, oxygen and experimental equipment. There were also a number of vital pieces of replacement parts for the Elektron generator that had failed some weeks before. The Elektron generator was the space station's primary source of oxygen and was a device that separated the oxygen from water by means of electrolysis. Up to this point the crew had been using Solid Fuel Oxygen Generating (SFOG) 'candles' to maintain their breathable atmosphere. There were also replacement candles aboard

the cargo spacecraft to replace the ones used. A new camera was also sent, to be used to photograph the thermal insulation tiles on the Orbiter spacecraft *Discovery* which was to dock with the ISS on 30 July. This was to be the first flight of the Space Shuttle since the *STS-107* disaster. Earlier problems discovered in the rudder speed braking system had caused all Space Shuttle orbiters to be grounded until the system had been replaced, but *Discovery* had been the first to be repaired. *STS-114* (*Discovery OV-103*) was launched on 26 July 2005, its mission to evaluate and test the new orbiter, but it was also tasked with taking much needed supplies to the ISS. This of course eased the pressure on the Russian Space Agency, as up to this point they had been the mainstay of the ISS and it had become a financial drain of their resources. During the mission there were a number of spacewalks to test and evaluate new inspection and repair techniques on the protective tiles on *Discovery*. Two days after launching, *Discovery* docked with the ISS, but as they approached, Sergei Krikalev and John Phillips used the digital cameras that had been delivered in the cargo spacecraft, and the high-powered 800mm and 400mm lenses, to photograph *Discovery*'s thermal protective tiles and key areas around its main and nose landing gear doors. All the images were down-linked to a team of specialists at NASA to analyse. After transferring supplies and equipment to the ISS, the orbiter *Discovery* undocked and returned to Earth. The landing site was moved to its secondary one of Edwards Air Force Base because of bad weather at the Kennedy Space Center. It was to be another year before the next Space Shuttle was to be launched.

The next cargo spacecraft, *Progress M-54*, was due to dock with the ISS on 24 August, but it was to be 10 September before it actually arrived. The unmanned cargo spacecraft blasted off the launch pad on 8 September 2005 on a Soyuz-U rocket, with supplies of food, oxygen and water for the two crew members, Sergei Krikalev and John Phillips, aboard the ISS. In addition to the normal supplies, the spacecraft also carried a number of new experiments, some additional tools and some spare parts. Among the spare parts was a new liquid unit for the Elektron device, which was the primary source of oxygen for the space station. Temporary repairs had been carried out, but the astronauts still had to rely on the reserve oxygen supplies and SFOG (Solid Fuel Oxygen Generator) 'candles'.

The spacecraft *Soyuz TMA-7* was launched on 3 October 2005 and included the third paying tourist, American Gregory Olsen. The multi-millionaire paid $20 million to the firm Space Adventures, who brokered flights to the space station. The resident ISS crew was replaced after their six-month mission by the *Expedition 12* crew of Bill McArthur (A), Valeri Tokarev (R) and Gregory Olsen (A). Unlike the previous two 'tourists', Gregory Olsen considered himself a scientist and carried out a number of experiments, including serving as a test subject for two human physiology studies for the ESA (European Space Agency), studying motion sickness and lower back pain. The latter was a condition that almost all astronauts suffered from with long duration gravity-free spaceflights. After ten days in space, Gregory Olsen returned to Earth on 10 October with the returning *Expedition 11* crew of Sergei Krikalev and John Phillips, landing in Kazakhstan.

The remaining crew of Bill McArthur and Valeri Tokarev started their six-month residency of the space station, and, as with all previous crews, most of their time would be keeping up with the maintenance and housekeeping duties, interspersed with various experiments. This invariably meant long working days and short periods of rest. One of the necessary tasks the crew had to do was to boost the space station into a higher orbit using the *Progress M-54* cargo spacecraft's engines. This had been docked at the aft end of the Zvezda module, and the idea was to carry out two burns, each to last 11 minutes and 40 seconds, that would raise the space station's orbit from its orbital peak of 220 miles to 224 miles. But when the engines were fired, they turned themselves off less than two minutes into the burn. Ground control discovered the problem was down to a fault in the data transmission which was quickly resolved.

On 7 November 2005, Bill McArthur and Valeri Tokarev donned American spacesuits and left the ISS to replace some of the tools on the exterior of the space station's outposts. The EVA took five and half hours during which time the space station was left empty, only the second time this had ever happened. The control of the station was left in the hands of the controllers back on Earth. It was also the first time that American EVA suits had been worn for two years. Up to this point the crews had had to rely on the Russian-built Orlan-M spacesuits because contamination to both the US-built Quest airlock and the EMU (Extravehicular Mobility Units) prevented the spacesuits from being used. Repairs had been carried out making the Quest airlock operational again. The EVA was also to place a camera and lights at the outer most edge of the space station. The camera, which weighed a couple of hundred pounds, was moved effortlessly by McArthur and Tokarev in the weightlessness of space. That completed, the two astronauts retraced their movements and removed a scientific device called the Floating Potential Probe that measured the electric potential of the space station as it travelled through the Earth's magnetic field. The device had not been working for some time and so was released and cast away to join other pieces of space junk. The probe burnt up some days later as it entered the Earth's atmosphere.

On the 10 November, the docked unmanned *Progress M-54* cargo spacecraft once again fired its engines and this time successfully moved the whole space station into a higher orbit. This was the second attempt to move the space station as the earlier one had failed after the engines had not fired long enough to carry out the manoeuvre. The new orbit, which was almost circular, placed the ISS in an orbit that ranged between 214 to 219 miles above the Earth.

The *Progress M-54* spacecraft was left at its docking port, whilst the *Progress M-55* cargo spacecraft docked at the nadir port on the Pirs docking module on 23 December 2005. The reasoning behind the retaining of *Progress M-54* was to enable the crew to stash more rubbish and unwanted materials into it, and later, on 3 March, to send the spacecraft into a decaying orbit to be burnt up in the atmosphere on re-entry.

The Pirs docking berth was the one usually reserved for the visiting crew-carrying Soyuz spacecraft and on 30 March 2006, a new crew was sent to the ISS aboard *Soyuz TMA-8*. Launched from Baikonur on a Soyuz-FG rocket, it docked in the nadir port

Pirs docking module separating from the ISS using a Progress cargo spacecraft. (NASA)

of the Zarya module on 1 April. The spacecraft carried the first Brazilian cosmonaut, Marcos Pontes. After docking and the usual safety checks had been carried out, the new crew of Pavel Vinograd (R), Marcos Pontes (Brazil), Jeffrey Williams (A) and the existing two crew members settled down to their tasks. The *TMA-7* crew had to finish their experiments in preparation for leaving and the *TMA-8* crew had to acclimatise themselves. On 7 April the *TMA-7* returning crew of Bill McArthur (A), Valery Tokarev (R) and Marcos Pontes (Brazil) got into their spacecraft and headed back to Earth. They landed near Arkaylk, Kazakhstan the following day.

After settling in to life aboard the space station, the two resident crew members, wearing Russian Orlan-MK spacesuits, carried out their first spacewalk, which lasted over six hours. Exiting through the Pirs module, they repaired a vent connected to the station's Elektron oxygen-producing unit by replacing a nozzle and the neck of the valve. This was a very important piece of equipment, as the Elektron breaks down water into its two components – hydrogen and oxygen. The oxygen is used in the space station, whilst the hydrogen was vented into space. They also replaced a camera on the orbiting rail car system amongst a number of other minor repairs and examinations. A number of external experiments were also removed from the Zvezda module to be examined inside the space station. The longest and most difficult part of the spacewalk, which was conducted by both the crew members, was to replace the malfunctioning camera

Soyuz TMA-8 crew. Top to bottom: Jeffrey Williams (A), Marcos Pontes (Braz) and Pavel Vinogradov (R). (NASA, Bill Ingalls)

on the Mobile Base System that operated the Canadarm 2 robotic arm that travelled along the station's main truss. With this completed the two men re-entered the ISS via the Pirs module for a well-earned rest. These long duration EVAs, although being carried out in a weightless environment, were extremely exhausting both physically and mentally. In the meantime, the cargo spacecraft *Progress M-56* had been launched on 24 April and docked in the aft port of the Zvezda module of the ISS two days later, bringing more supplies and equipment to the resident crew. It was to stay connected to the space station until 19 September, when during this period it was filled with the usual unwanted materials and waste, then placed into a decaying orbit to be burnt up over the Pacific Ocean as it entered the Earth's atmosphere.

Another crew member joined the ISS when the Space Shuttle *STS-121* (*Discovery OV-103*) arrived with ESA astronaut Thomas Reiter (Ger) on 6 July 2006. He was later to take part, with Jeffrey Williams, in the second spacewalk that lasted five hours and consisted of installing and recovering experiments and moving various pieces of equipment, dictated to by mission control. This was the first time in more than three years that the ISS had had a third crew member and was a great help in helping the other two crew members suit up for their EVAs.

A potentially serious incident arose when the oxygen generator overheated, discharging a toxic irritant. The crew immediately donned masks and gloves and cleaned up the spillage with towels. A charcoal filter was then placed close to the generator to take the toxic irritant out of the air. Within three hours all trace of the irritant had been removed and tests showed the air inside the ISS to be clean. This was the first serious incident on the ISS, and although pales into insignificance compared to incidents on *Apollo 13* and *Mir*, it was a salutatory reminder about the risks of living in space. The crew later openly admitted that they were very concerned at the time, but their emergency training had showed its worth.

The next cargo spacecraft, *Progress M-57*, arrived on 24 June with fresh supplies and equipment. Launched on a Soyuz-U companion rocket, it docked in the nadir port of the Pirs module on the ISS. It was to stay docked until the 16 January 2007 when it was undocked to make way for *Progress M-58*. *Progress M-57* entered the Earth's atmosphere the following day and was burnt up over the Pacific Ocean.

The Space Shuttle *STS-115* (*Atlantis OV-104*) arrived on 10 September 2006 to fit a P3/P4 truss structure to the space station. This addition to the ISS was designed to help expand the space station by attaching modules to it.

The next relief crew to the ISS was launched a week later on 18 September 2006, when *Soyuz TMA-9* (*Expedition 14*) on a Soyuz-FG rocket blasted off the launch pad at Baikonur. The crew were Mikhail Tyurin (R), Michael E. Lopez-Alegria (A) and tourist Anousheh Ansari (Iran/A). Originally a Japanese space tourist, Daisuke Enomoto, had been scheduled to be on the spaceflight, but he was deemed to be unfit due to medical reasons and so his place was taken by Anoushen Ansan, an electrical engineer. When they docked with the ISS in the aft port of the Zvezda module, there were for the first time twelve astronauts/cosmonauts in space at the same time: three on the space station, three on *TMA-9* and six on *Atlantis*. The orbiter was to undock

The crew of *Soyuz TMA-9*. L- R: Anousheh Hansan (A), Mikhail Tyurin (R) and Michael Lopez-Alegria (A). (NASA)

on 17 September and return to Earth. Eleven days after arriving, Anousheh Ansari returned to Earth on 29 September with the crew of *TMA-8*, Jeffrey Williams and Pavel Vinograd, landing in Kazakhstan. On 10 October, *Soyuz TMA-9* undocked from the Zvezda module and re-docked in the nadir port of the Zarya module to make way for the next Progress cargo supply craft.

Progress M-58 was launched from Baikonur on a Soyuz-U rocket on 23 October 2006 with more supplies and experiments for the resident crew. The cargo spacecraft docked in the aft port of the Zvezda module on 26 October, but there was some initial concerns after an indicator indicated that one of the docking antennas on the Kurs docking system had failed to retract. After a safe docking it turned out to have been a false reading. The ISS now had two Progress supply craft attached.

The first of the *Expedition 15* crew, Sunita Williams (A), arrived on the ISS on 11 December 2006 aboard *STS-116* (*Discovery OV-103*) that had been launched on 9 December. She was to join the *Expedition 14* crew until the arrival of the *Expedition 15* crew on *Soyuz TMA-10*. After docking in the PMA-2 port, the crew of *Discovery* delivered a P5 truss segment and carried out a rewiring on the space station's power system. They then collected Thomas Reiter (ESA), who was being replaced by Sunita Williams, undocked on 19 December and returned to Earth.

More supplies and equipment arrived in January with the arrival of *Progress M-59* which was launched from Baikonur on 18 January 2007. After a two-day flight the

cargo spacecraft docked in the nadir port of the Pirs module, where it stayed after being unloaded for the next six months during which time it was filled with unwanted materials and waste. It wasn't until 1 August before it was undocked and placed into a decaying orbit to burn up on re-entry over the Pacific Ocean.

The ISS became slightly crowded on 9 April with the arrival of *Soyuz TMA-10*. Launched from Baikonur on a Soyuz-FG rocket on 7 April 2007, the spacecraft docked in the nadir port of the Zarya module to deliver a fresh crew and some supplies to the ISS. On board with the two Russian crew members, Oleg Kotov and Fyodor Yurchikhin, was American space tourist Charles Simonyi, a Microsoft executive, who had arranged for the ten-day visit to the ISS for an undisclosed sum. After passing the physical and completing the six months of mandatory training for visitors, he was assigned to *TMA-10*. During his ten-day visit, Simonyi, a qualified radio ham, broadcast on amateur radio to fellow enthusiasts around the world. In the meantime the resident crew and the new crew continued with their experiments and housekeeping duties. With these completed, Charles Simonyi, Mikhail Tyurin and Michael Lopez-Alegria climbed into the *Soyuz TMA-9* spacecraft and returned to Earth, landing over 400km away from the intended landing area of Arkaylk, Kazakhstan. There was some concern during the re-entry when, whilst entering the atmosphere, the spacecraft suddenly transitioned to a ballistic re-entry. (A ballistic re-entry is a back-up re-entry that automatically takes over if there is a failure during normal re-entry.) It was said later that a cable in the spacecraft's control panel that was connected to the Soyuz descent equipment had been damaged, causing the explosive bolts situated in the struts between the re-entry module and the service module, to fail to explode. It turned out later that this was not true, the problem had been caused due to the connecting struts melting in the heat of the re-entry, causing a change in the spacecraft's orientation, which in turn activated the back-up system. This was not the first time this had happened. A similar incident had occurred on *Soyuz 5* and on *Soyuz TMA-1* fortunately without any further problems. Because the re-entry module was, and still is, covered in thermal insulation, the struts melted before any damage was caused to the capsule itself. The spacecraft landed safely near Arkalyk, Kazakhstan.

The next visitor to the ISS was the cargo spacecraft *Progress M-60*. It was launched on a Soyuz-U companion rocket on 12 May 2007, with more supplies and equipment for the ISS. After the normal two-day journey, the cargo spacecraft docked in the aft port of the Zvezda module, where it stayed until 19 September during which time it was filled with waste and other unwanted materials. The Space Shuttle, *STS-117* (*Atlantis OV-104*), was launched on 8 June 2007 to deliver ISS crew member Clayton Andersen and a truss. He was to join the resident crew aboard the space station and replace Sunita Williams who, after completing all her tasks and experiments, returned to Earth on *STS-117*. Another cargo spacecraft, *Progress M-61*, was launched from Baikonur on 2 August 2007 carrying supplies to the ISS. After a two-day flight, the spacecraft docked in the nadir port of the Pirs module. It was later moved to the aft port of the Zvezda module to make way for the arrival of *Soyuz TMA-11*.

Progress M-60 was undocked from the Zvezda module on 19 September and placed into a decaying orbit until 25 September when it was burnt up on re-entry over the Pacific Ocean. *Progress M-61* remained docked until 22 December during which time it too was filled with discarded materials and waste and undocked from the space station to drift in free orbit. It wasn't until 22 January 2008 before it was finally placed into a decaying orbit to be burnt up on re-entry.

Launched on 10 October 2007 on a Soyuz FG rocket, *Soyuz TMA-11* carried the *Expedition 16* relief crew of Yuri Malenchenko, Peggy Whitson and Sheikh Muszaphr Shukor to the ISS. Although initially classified as a space tourist, the Malaysian cosmonaut had completed the full training required for all professional members of the space crews going to the ISS and was considered a member of the cosmonaut corps by the other cosmonauts and not just a paying tourist. In fact it was the Russian Federation that footed the bill for his training, after the Malaysian government had completed a multi-billion dollar deal for the purchase of fighter jets. After docking in the Zarya module and completing all the safety checks, the three cosmonauts entered the ISS to relieve the resident crew of Fydor Yurchikhin, Clayton Andersen and Oleg Kotov. After the welcoming ceremony, both sets of crews set about their tasks of experiments and housekeeping duties. On 21 October, the *Expedition 15* crew of Fydor Yurchikhin, Oleg Kotov and Sheikh Muszaphr Shukor handed over command of the space station, said their farewells and returned to Earth in *Soyuz TMA-10*, landing near Arkaylk, Kazakhstan. The American astronaut Clayton Andersen remained aboard with Peggy Whitson, who was now the commander, and Yuri Malenchenko. Clayton Anderson returned to Earth aboard *STS-120* after it had delivered the Harmony node and replaced him with Daniel Tani (A).

The cargo spacecraft *Progress M-62* was launched on a Soyuz-U rocket on 22 December 2007 with supplies and experiments, and no doubt, a few Christmas surprises for the crew. After the two-day trip to the space station, the cargo spacecraft docked with the nadir port of the Pirs module, where it stayed until 4 February 2008 whilst being used as a trash can for discarded waste and materials before being placed into a decaying orbit to be burnt up on re-entry. One day after *Progress M-62* undocked from the Pirs module, *Progress M-63* arrived and docked in the nadir port of the Pirs module with more supplies. Once emptied, it too was filled over the next couple of months with the usual waste and unwanted materials, then the cargo spacecraft was undocked on 7 April 2008 and, unusually, was placed straight into an orbit that took it directly into the Earth's atmosphere. The following day, *Soyuz TMA-12* (*Expedition 17*) was launched from Baikonur on a Soyuz-FG rocket, with the ISS relief crew of Alexsandr Volkov, Oleg Kononenko and Yi So-yeon (Kor) on board. Yi So-yeon was listed as a space tourist after the South Korean government had paid the Russian Federation 25 million US dollars. During her eight days aboard the ISS, she carried out a number of biotechnological experiments for the South Korean government. Originally she had been the back-up for another South Korean, Ko San, but during his training he had breached one of the very strict regulations by removing books from the centre and, by doing so, forfeited the opportunity of being the first

South Korean to fly in space. Yi So-yeon was to leave the space station on 19 April together with the returning ISS crew of Peggy Whitson and Yuri Malenchenko aboard *Soyuz TMA-11*. Problems arose as they approached re-entry when the aft section of the Soyuz spacecraft failed to separate, causing it to re-enter in a reverse position. This meant that the forward hatch was taking the full force of the heat generated by the speed of the spacecraft as it re-entered the Earth's atmosphere. This was not the first time this had happened and fortunately, as with *Soyuz 5* in 1970 and *Soyuz TMA-10* the previous year, the connections melted causing the spacecraft to right itself and allowing the heatshield to absorb the heat. This was known as a 'ballistic re-entry' with the spacecraft entering at a very steep angle instead of a shallow one, causing *Soyuz TMA-11* to land hard and 470 kilometres short of the estimated landing area just outside of Arkaylk, Kazahkstan. Yuri Malenchenko and Peggy Whitson suffered a few minor injuries, but nothing serious. However on impact Yi So-yeon was hit by Whitson's personal effects bag and required treatment for minor neck and spine injuries. The Russians had kept the problem of *Soyuz TMA-10* a secret until it happened again on the *Soyuz TMA-11*, this time with a NASA astronaut on board. This infuriated NASA, claiming that the Commission of Inquiry had misled them and demanded further investigation. The Russians responded by carrying out a special EVA activity on the ISS to check the docked Soyuz *TMA-12* and its explosive bolts in its connection struts. The results of the enquiry are not known.

It wasn't until 14 May 2008 that the next cargo spacecraft, *Progress M-64*, was launched on a Soyuz-U rocket with fresh supplies for the crew on the ISS. The spacecraft docked in the aft port of the Zarya module on 16 May, where it remained until 1 September when it was undocked. During this period, it was being filled with unwanted materials and rubbish. The cargo spacecraft remained in a free orbit until 8 September when it was placed in a decaying orbit and entered the Earth's atmosphere to be burnt up on re-entry over the Pacific Ocean. Ten days later, on 10 September 2008, *Progress M-65* was launched on a Soyuz-U rocket, the unmanned cargo spacecraft docking in the aft port of the Zvezda module on 13 September. After being unloaded it too was re-filled with unwanted materials and rubbish over the next couple of months. It was finally undocked on 15 November, but it wasn't until 8 December that it went into a decaying orbit. It was burnt up as it entered the Earth's atmosphere over the Pacific Ocean. In the meantime another crew had arrived on the ISS, when on 12 October 2008, *Soyuz TMA-13* (*Expedition 18*) was launched on a Soyuz-FG rocket and docked two days later in the nadir port of the Zarya module. The crew consisted of Yuri Lonchekov (R), Michael Finke (A) and Richard Garriot (A/UK). Richard Garriot was the son of the former NASA astronaut Owen Garriot (*Skylab 3*, *STS-9*) and was labelled a space tourist because he was privately funded for the trip. He returned to Earth ten days later on *Soyuz TMA-12* with the returning members of *Expedition 17*: Alexsandr Volkov and Oleg Kononenko. Owen Garriot was present at the launch and the landing of his son's mission, setting a first with a father and son having both flown in space.

A new type of cargo spacecraft was launched on a Soyuz-U rocket on 26 November 2008, the *Progress M-01M*, carrying supplies to the ISS. This new cargo spacecraft

Soyuz TMA-14 approaching the ISS. (NASA)

featured a digital flight computer and telemetry system in place of the analogous system that had been fitted in the earlier models. After a two-day flight, the spacecraft docked in the nadir port of the Pirs module where it stayed for the next ten days. After being unloaded it was filled with waste and unwanted materials and left docked for the next three months. On the 6 February 2009 it was undocked and placed into a decaying orbit where on 8 February, it burnt up on re-entry. It was almost three months, until 10 February 2009, when the next cargo spacecraft, *Progress M-66*, was launched on a Soyuz-U rocket. The spacecraft docked in the nadir port of the Pirs module on 13 February and was to remain there until 6 May when it was undocked and placed into a decaying orbit. For the next twelve days it orbited the Earth before it entered the Earth's atmosphere on 18 May and was burnt up on re-entry over the Pacific Ocean.

The American Space Shuttle *STS-119* (*Discovery OV-103*) arrived on 17 March 2009 to deliver the S6 Solar array to the ISS, which completed the Integral Truss Structure. In addition to delivering a number of experiments, the shuttle also delivered another crew member onto the station, flight engineer Koichi Wakata from Japan. He was to join the oncoming *Expedition 19* crew as a resident crew member. The next resident crew to the ISS, *Soyuz TMA-14* (*Expedition 19*), was launched on 26 March 2009 on a Soyuz-FG rocket. The crew of Gennady Padalka (R) and Michael Barratt (A) were joined by Charles Simonyi (A), who was a space tourist on his second flight; his first

Soyuz TMA-15 on its way to the launch pad aboard a specially-constructed railway carriage. (Roscosmos)

was on *Soyuz TMA-9*, he having paid for the trip through the brokerage company Space Adventures. The Soyuz spacecraft docked with the ISS on 28 March in the aft port of the Zvezda module.

With all the safety checks completed, the oncoming crew opened the hatch to be met by a welcoming resident crew. Japanese flight engineer Koichi Wakata was already on board, having been delivered on 17 March by the Space Shuttle *STS-119*. On 2 July, *Soyuz TMA-14* was moved from the Zvezda module to the nadir port of the Pirs module to make way for the arrival of the *Progress M-02M* cargo spacecraft. After completing his experiments, Charles Simonyi returned with the relieved crew of Yuri Lonchekov and Michael Finke (*Expedition 18*) on 21 April. There was a problem however during the re-entry of *Soyuz TMA-13* causing the spacecraft to enter into a 'ballistic re-entry' of 9g instead of the normal 4g, resulting in a hard landing some 338 kilometres short of the target landing area. Fortunately no-one was injured. No explanation was given for landing short of the expected area.

Unlike previous ISS crews, a morale problem suddenly manifested itself when Gennady Padalka was made aware of issues back on Earth regarding the use of exercise equipment and the toilet facilities being shared with the Americans on the space station. Padalka claimed that his request to use the American-owned exercise equipment had been turned down by ground control, so it was decided that in future the Russian and American members of the crew would not only use their own equipment and toilet facilities, but they were not to share rations. However Gennady Padalka said all these petty squabbles had started when Russia agreed to take paid tourists to the ISS. He continued saying that up to this point all the cosmonauts and astronauts, who were all intelligent, educated professionals, had worked together harmoniously and as far as he was concerned would continue to do so. No mention was made of the facilities concerning the Japanese member of the crew, Koichi Wakata, so one assumes that as the Americans had brought him, so if there was a problem, it was probably the American facilities that he used. It appears for some unknown reason that the mistrust between the two super powers had raised its ugly head once again. As far as it is known the subject matter was never raised again.

Progress M-02M was launched on 7 May 2009 on a Soyuz-U rocket and docked in the Pirs module on 12 May. After being unloaded the cargo spacecraft was refilled with waste and unwanted materials, including two Orlan-MK spacesuits, and on the 30 June, it was undocked and allowed to free flight until 12 July when it performed a second docking attempt to test the docking system in preparation for the arrival of Mini-Research Module-2 in the Zvezda module. The cargo spacecraft approached within 33–39ft (10–12 metres) then backed away. The following day it went into a decaying orbit and was burnt up on re-entry.

In the meantime *Soyuz TMA-15* (*Expedition 20*) had been launched on 27 May 2009 on a Soyuz FG companion rocket carrying the relief crew of Roman Romanenko (R), Frank De Winne (Belg/ESA) and Robert Thirsk (Can). This was another of the rare full international crews to visit the ISS. After a two-day flight, the spacecraft docked in the nadir port of the Zarya module on 29 May. The relief crew joined with the resident crew of Gennady Padalka, Michael Barratt and Koichi Wakata, making up a six-manned crew of the International Space Station. To prepare for the arrival of the Russian Mini-Research Module-2 (MRM-2), a number of modifications had to be made to the Zenith port on the Zvezda module, so Gennady Padalka and Michael Barratt carried out a five-hour EVA to start the work. To complete the task, a further spacewalk was required, only this time it was an Internal Vehicular Activity (IVA) and was inside the ISS or, to be more precise, to be carried out inside the depressurised Zvezda module. There the two crew members removed one of the hatches and replaced it with a docking cone in preparation of the arrival of the Poisk module (MRM-2). The whole IVA mission took just twelve minutes, one of the shortest spacewalks in the history of the ISS. With this task completed the spacecraft *Soyuz TMA-14* was then undocked on 2 July from the Zvezda module and moved to the Pirs module to make way for the arrival of the Progress cargo spacecraft *M-67*.

The next manned flight to the ISS was the Space Shuttle *STS-127*, which was launched on 15 July 2009 to deliver and install the final components of the Japanese Experimental Module and deliver another ISS crew member, Timothy Kopra (A), to replace Koichi Wakata (Jpn). *Endeavour* docked in the PMA-2 (Harmony) port on 17 July creating a milestone in the history of space exploration as there were now thirteen astronauts/cosmonauts in space together for the very first time. *Endeavour* undocked from the ISS on 28 July with Koichi Wakata on board.

Another cargo spacecraft arrived on 26 July 2009, *Progress M-67*. Launched on 24 July on a Soyuz-U rocket, the cargo spacecraft docked in the aft port of the Zvezda module. After being unloaded it was refilled over the next seven weeks with waste and unwanted materials. It was undocked on 21 September, but it wasn't until

27 September before it went into a decaying orbit and burnt up over the Pacific Ocean.

Another ISS crew member was delivered to the ISS by the Space Shuttle *STS-128 (Discovery OV-103)* on 1 September. American astronaut Nicole Stott was to replace Timothy Kopra, who then joined the crew of *STS-128* and returned to Earth. Nicole Stott was to return to Earth aboard *STS-129* and was to be the last ISS crew member to return in the Space Shuttle.

One month later on 30 September 2009, *Soyuz TMA-16* (*Expedition 21*) was launched on a Soyuz FG companion rocket, to deliver the next relief crew for the ISS, Maksim Surayev (R), Jeffrey Williams (A) and space tourist Guy Liberté (Can), and docked in the nadir port of the Zvezda module. After the usual safety checks, the crew cracked the hatch to be welcomed by the *Expedition 19* crew. After the formalities both crews got down to the duties required to maintain the space station and of course their own projects. On 11 October 2009 Gennady Padalka formally handed command of the ISS to Maksim Surayev; he and his fellow crew members Michael Barratt and Guy Liberté (*TMA-16*) then said their farewells, climbed into *Soyuz TMA-14* and returned to Earth, landing near Arkaylk, Kazakhstan.

The next cargo spacecraft, *Progress M-03M*, was launched on a Soyuz-U rocket on 15 October 2009 and docked on 18 October in the Pirs module. The cargo spacecraft delivered 1,750lb (790kg) of dry cargo and, after being unloaded, it was refilled over the next six months with about the same amount of waste and unwanted materials. Another cargo spacecraft, *Progress M-MIM2*, was launched on 10 November on a Soyuz-U carrier rocket and docked with the ISS on 12 November in the Zenith port of the Zvezda module to deliver another module, the Poisk. Once unloaded, over the next few weeks the now empty cargo spacecraft was filled with unwanted waste. On 8 December it was undocked and placed in a decaying orbit to burn up over the Pacific Ocean on re-entry.

The *Soyuz TMA-15* crew of Roman Romanenko, Frank De Winne and Robert Thirsk said their farewells and returned to Earth on 1 December 2009, landing at Arkalyk, Kazakhstan.

Soyuz TMA-17 (*Expedition 22*) was launched on 20 December 2009 on a Soyuz-FG companion rocket with the relief crew of Oleg Kotov (R), Soichi Nogichi (Jpn) and Timothy Creamer (A), and docked in the nadir port of the Zarya module. This was Soichi Nogichi's second visit to the ISS, his first was as a crew member aboard the Space Shuttle *STS-114* (*Discovery OV-103*) which had delivered supplies to the resident crew on the space station. The *Soyuz TMA-16* spacecraft was undocked from the Zvezda module on 21 January and re-docked in the Poisk port to make way for the arrival of the next spacecraft. On 3 February 2010, *Progress M-04M* was launched on a Soyuz-U rocket, then two days later it docked in the nadir port of the Zarya module. This docking marked the moment that four Russian spacecraft, *Progress M-04M*, *Soyuz TMA-17*, *Soyuz TMA-16* and *Progress M-03M*, were docked with the ISS at the same time. With the arrival of the relief crew and the handing over command of the space station completed, the relieved crew of Maksim Surayev and Jeffrey Williams returned to

Earth aboard *Soyuz TMA-16* on 18 March, landing in Kazakhstan. *Progress M-03M* undocked on 22 April 2010 and entered the Earth's atmosphere five days later to be burnt up on re-entry over the Pacific Ocean.

Soyuz TMA-17 was later moved to the aft port of the Zvezda module to make way for the arrival of the Rassvet module that was to be delivered later that month by the Space Shuttle (*Atlantis OV-104*) on *STS-132*. In the meantime *STS-130* (*Endeavour OV-105*) was launched on 8 February 2010 to deliver the Tranquillity module and Cupola to the ISS. This was a robotically controlled station with seven windows, six around the sides and one in the centre, giving a 360-degree view around the station. With the work completed *Endeavour* undocked and returned to Earth on 22 February.

Soyuz TMA-18 (*Expedition 23*) was launched on 2 April 2010 on a Soyuz-FG rocket with the relief crew of Alexsandr Skvortsov, (R), Mikhail Komienko (R) and Tracy Caldwell Dyson (A), the spacecraft docking in the Poisk module on 4 April. Prior to opening the hatch, the new crew carried out the necessary leak checks then removed their Sokol space suits and gloves and prepared them for drying. It was almost two hours before the hatch was finally opened for them to be welcomed by the resident crew. During their stay aboard the ISS they encountered a few minor problems.

The next cargo spacecraft, *Progress M-05M*, was launched on 28 April 2010 on a Soyuz-U rocket. Two days later on 1 May, it docked with the Pirs module. In addition to the usual food, water, propellants and research materials, this trip included a number of luxuries, such as confectionery, books and a number of movies. After being unloaded, the spacecraft was refilled with unwanted materials and waste, but it remained docked until 25 October. Just prior to undocking, the hatch between the space station and the Progress spacecraft was checked for leaks. After undocking, the spacecraft was placed into a lower orbit where it orbited a safe distance from the space station, enabling Russian scientists to conduct a number of geophysical experiments before it entered the Earth's atmosphere and was burnt up over the Pacific Ocean on 15 November 2010. *Progress M-04M* remained docked until 10 May before being placed into a free orbit until the 1 July when it finally entered the Earth's orbit and was burnt up on re-entry over the Pacific Ocean. The *Soyuz TMA-17* spacecraft, which had originally docked in the nadir port on the Zarya module, was moved on 12 May to the Zvezda module to make way for the arrival of the next Progress cargo spacecraft.

The Americans were the next to visit the International Space Station with *STS-132* (*Atlantis OV-104*). It was launched on 14 May 2010 to deliver the Russian MRM-01 Rassvet module to the space station. After safely delivering the module, *Atlantis* undocked from the ISS to carry out a series of experiments for the Americans.

With the handover to *Soyuz TMA-18* completed, preparations were made for the relieved crew of Oleg Kotov (R), Soichi Nogichi (J), Timothy Creamer (A) (*TMA-17*) to return to Earth. On 26 May, in order to ensure perfect conditions for their re-entry, the orbit of the space station was lowered by 1.5 kilometres using the engines on the *Progress M-05M* cargo spacecraft. The *Soyuz TMA-17* crew undocked from the ISS on 1 June and headed for Earth, landing four hours later on the steppes of Kazakhstan.

Two weeks later *Soyuz TMA-19* was launched to deliver the *Expedition 24* crew to the ISS. The crew, Fyodor Yurchikin (R), Shannon Walker (A) and Douglas Wheelock (A), docked in the aft port of the Zvezda module on 18 June after an uneventful flight. The arrival of *TMA-19* created a six-man crew on the ISS. On 28 June the three crew members climbed back into their spacecraft and Fyodor Yurchikin manually undocked to about forty metres away from the space station and then re-docked in the newly installed Rassvet module twenty-five minutes later. During their mission, the crew members worked on more than 120 microgravity experiments in human research, biology and biotechnology, physical and materials sciences, technology development, and Earth and space sciences. An emergency arose after more than half of the space station's cooling system shut down, resulting in three unplanned spacewalks being required to fix the problem. Douglas Wheelock and Tracy Caldwell carried out the spacewalks to replace the faulty pump module that caused the shutdown and were able to fully restore the cooling system. The *Expedition 24* crew returned to Earth on 26 November 2010, landing near Arkalyk, Kazakhstan.

The next visitor to the space station was *Progress M-06M*. Launched on 30 June 2010 on a Soyuz-U companion rocket, it docked on 4 July in the aft port of the Zvezda module. The first docking attempt on 2 July failed when the unmanned spacecraft received an abort signal from the Kurs automated rendezvous system. The cargo spacecraft then went into orbit, bypassing the space station, then came around for a second attempt on 4 July: this time it was successful. After being unloaded and refilled with waste and unwanted materials, it was undocked on 31 August and placed into a decaying orbit to be burnt up over Pacific Ocean on re-entry.

Another cargo spacecraft, *Progress M-07M*, was launched on 10 September on a Soyuz-U companion rocket and docked in the aft port of the Zvezda module on 12 September. This time the Kurs automatic rendezvous system guided the cargo spacecraft into a flawless docking. *Progress M-07M* carried 2,504lb (1,136kg) of dry cargo, 2,470lb (1,120kg) of fuel, 108lb (49kg) of oxygen and 460lb (210kg) of water. This was in addition to the household equipment required for maintenance and repairs, food, fresh clothes and medical equipment that was needed. There were also cameras that were to be fitted into the Zarya, Rassvet and Pirs modules. Once unloaded, a series of manoeuvres were carried out to raise the orbit of the International Space Station using the Progress engines. The first, on 15 September, used eight of the spacecraft's attitude control thrusters which raised the space station by 1.2 miles (2 kilometres) to 221 miles (356 kilometres). This was done in preparation for the undocking of *Soyuz TMA-18* on 25 September and the docking of *Soyuz TMA-01M*, due on 10 October 2010.

On 23 September 2010, after handing over command of the ISS to Fydor Yurchikhin, Alexsandr Skvortsov, Mikhail Komienko and Tracy Caldwell Dyson entered their spacecraft in preparation to return to Earth, but on closing the hatch, the crew experienced problems getting a tight seal. They opened the hatch for a quick inspection, but could find nothing wrong. They re-sealed the hatch, but Fydor Yurchikhin, who had arrived on *Expedition 24* and was working inside the space station monitoring the undocking, announced that he was not getting confirmation

fraternity. The arrival of *Progress M-11M* on 23 June 2011 heralded the arrival of fresh supplies and more equipment for the resident crews aboard the ISS. Launched on the 21 June on a Soyuz-U companion rocket, the cargo spacecraft docked on the aft port of the Zvezda module. After unloading the supplies, the crew then reloaded the cargo spacecraft with unwanted materials and waste. On 1 July the four attitude thrusters on the Progress cargo spacecraft were activated to move the space station into a higher orbit, raising it by 3.5 kilometres. The reason this was done was to make way for the Space Shuttle *STS-135* mission to the ISS. This was to be the last mission for the Space Shuttle as the program was to be shut down, leaving all deliveries of crews and supplies to the ISS squarely in the hands of the Russians. This was a costly and tremendous responsibility to be placed on one nation. On 8 July 2011, *STS-135 Atlantis* (*OV-104*) was launched from the Kennedy Space Center, followed two days later with a smooth docking with the station's Pressurized Mating Adapter-2 (PMA) at the Harmony Node-2 module. On opening the hatches, the crew were welcomed by the six other crew members on the space station where they held a brief welcoming ceremony. After unloading a large number of supplies for the station, the Shuttle crew carried out a series of experiments and exercises before undocking on 20 July and returning to Earth the following day. Just prior to leaving the Shuttle crew presented the ISS crew with a small American flag that had been flown on the first Shuttle flight (STS-1) on 12 April 1981.

American flag flown on first shuttle flight, left on the ISS by *STS-135* to commemorate the first and last Space Shuttle flights. (NASA)

Another cargo spacecraft, *Progress M-12M*, was launched on 24 August 2011, but failed to reach orbit after the third stage of the Soyuz-U rocket cut out prematurely and crashed in the Choisk Region of Russia's Altai Republic. The failure of the *Progress M-12M* rocket could have created serious problems in supplying the crews on the space station, but Mission Control confirmed that there were plenty of supplies aboard, so there was no immediate problem. The thought of being stranded on the space station was not a pleasant one and plans had to be put in place for such an emergency. One of the options that was considered was leaving one of Soyuz TMA spacecraft *in situ* as a 'lifeboat'. In the meantime the *Soyuz TMA-21* crew, who had been preparing to leave the ISS on 8 September, were told that their departure had been put back because of the loss of the Progress cargo spacecraft. It wasn't until 16 September that *Soyuz TMA-21* finally undocked and headed back to Earth, but their problems continued, when as they started to deorbit they lost all communications with ground control, causing some worrying moments for the controllers. However the descent and the landing went as planned and the spacecraft touched down in Kazakhstan. No reason was ever given for the communications blackout. It wasn't until 30 October before the next cargo spacecraft was launched, *Progress M-13M*, on a Soyuz-U companion rocket and, on arriving two days later, docked in the nadir port of the Pirs module. This was a very welcoming sight for the crew of the space station as they were starting to run low on a number of necessities and commodities, but it was also a worrying time all round making everyone aware of the dangers of spaceflight. After unloading all the supplies, the crew refilled the cargo spacecraft with the usual waste and unwanted materials over the next few months and then closed the hatch. The Progress cargo spacecraft remained docked until 23 January 2012 before it was undocked and placed in a decaying orbit. It re-entered the Earth's atmosphere two days later and was burnt up over the Pacific Ocean. The successful launch and docking of *Progress M-13M* seemed to allay some of the fears about future flights, but the concerns would always remain.

Launched from Baikonur on 14 November 2011 on a Soyuz-U rocket, *Soyuz TMA-22* arrived after a two-day flight, docked in the MRM-2 Poisk module, safely delivering the *Expedition 29* crew of Anton Shkaplerov (R), Anatilo Ivanishin (R) and Daniel Burbank (A) to the ISS. After checking all the seals to ensure that there were no leaks, the relief crew cracked the hatch and entered the space station to be welcomed by the resident crew. The Americans were now paying for a seat on the Soyuz spacecraft so as to continue with their research and maintain a presence on the ISS. The launch had been scheduled for 30 September, but because of the failure of the *Progress M-12M* rocket, it had been put on hold. This was the first manned flight to the ISS since the American Space Shuttle program had ended. There had been some hesitancy regarding the launch on the day, because of the high winds and blizzard conditions at Baikonur, but it was considered within the parameters of safety and so the spacecraft lifted off the launch pad on schedule and into space. The Russian spacecraft were now the only means available to take fresh crews and supplies to the International Space Station and this was causing some financial worries. These disposable Progress

Soyuz TMA-3 being rolled out to the launch pad. (RKS Energia)

spacecraft cost millions to make as did the manned versions, which placed a huge financial burden on the Russians. In the meantime the *Soyuz TMA-02M* crew of Sergey Volkov, Satoshi Furukawa and Michael Fossum undocked from the space station on 21 November and landed safely on 22 November near Arkaylk, Kazakhstan.

One month later, on 21 December, *Soyuz TMA-03M* was launched on a Soyuz-FG companion rocket to deliver the *Expedition 30* crew of Oleg Kononenko (R), Andrei Kuipers (ESA) and Donald Petit (A) to the ISS. They were to spend Christmas and New Year with the resident *Expedition 29* crew before the latter departed on 21 April 2012, landing the following day near Dzhezkazgan, Kazakhstan.

Most of the research and experiments that were carried out on the space station were concerned with cardiovascular and physiology investigations pertaining to weightlessness and working in a space environment. There were however certain investigations by the different countries that were carried out under a blanket of security, so little is known about these. In the meantime the next cargo spacecraft, *Progress M-14M* on a Soyuz-U companion rocket, had lifted off the launch pad on 25 January 2012 and after a three-day flight, the cargo spacecraft docked in the nadir port of the Pirs module with much needed supplies and materials for the crews. Like all the previous Progress spacecraft it was refilled with unwanted materials and waste over the next few weeks. It was undocked on 19 April 2012 and placed in a decaying orbit. For the next nine days it orbited the Earth before finally entering the Earth's atmosphere and burning up over the Pacific Ocean.

Another cargo spacecraft, *Progress M-15M*, was launched on a Soyuz-U companion rocket on 20 April 2012 and after a two-day flight, docked in the nadir port of the Pirs module using the Kurs automated rendezvous and docking system. After the supplies had been unloaded, the spacecraft was filled with unwanted materials and waste. Amongst the cargo there was a birthday present for Gennadi Padalka, who was scheduled to arrive on 15 May, to celebrate his 54th birthday on 21 June 2012. In the meantime the next Expedition crew arrived to relieve the resident crew. *Soyuz TMA-04M* was launched on 15 May 2012 to deliver the *Expedition 31* crew of Gennady Padalka (R), Sergei Revin (R) and Joseph Acaba (A) to the ISS. After the usual safety checks had been completed, they cracked the hatch to be greeted by the resident crew. With the welcoming ceremony over the crews settled down to their tasks and the usual housekeeping duties. Two weeks later, on 1 June, the *TMA-03* crew, after handing over command of the space station, climbed into their Soyuz spacecraft, undocked and landed in Kazakhstan the following day.

Six weeks later, on 15 July 2012, the resident crew were joined by the *Expedition 32* relief crew of Yuri Malenchenko (R), Sunita Williams (A) and Akihiko Hoshide (Jpn) in their *Soyuz TMA-05M* spacecraft, which had been launched atop a Soyuz-FG rocket. After docking in the nadir port of the Rassvet module using the Kurs automated rendezvous system, the crew completed their safety checks and joined their colleagues on the ISS. A new Kurs navigation antenna had been fitted earlier, so on 22 July it was decided to test it by undocking the *Progress M-15M* spacecraft and then carrying out a re-docking procedure using the antenna. The re-docking was to take place two days later on 24 July, but was aborted because the on-board equipment failed a self-test. At the time the ISS and the Progress spacecraft were flying 15 kilometres apart, and a new attempt to dock was re-scheduled for 28 July after it was realised that the ISS was awaiting the arrival of the Japanese *Kounoton 3* spacecraft on 27 July. The cause of the docking problem was discovered to have been the low temperature in the Progress spacecraft, so the temperature was increased and a second docking attempt made. This time it was successful. The Progress spacecraft remained docked to the ISS until 29 July when it was undocked to make way for the arrival of another Progress cargo spacecraft, but it continued to be used to conduct two more experiments, the Khloppushka and a Radar Progress, after which it went into a decaying orbit to be burnt up on re-entry over the Pacific Ocean.

The arrival of another cargo spacecraft, launched on a Soyuz-U rocket on 1 August 2012, was a welcome sight to the ISS crew. The cargo spacecraft, *Progress M-16M*, docked in the nadir port of the Pirs module the following day, which was an unusually quick arrival. This was the first time a fast approach method had been used, docking on just the fourth orbit. The reasoning behind this was to allow the transportation of critical biological experiments to the ISS. The success of these tests allowed the method to be used on the transportation of crews to the ISS, thus reducing the fatigue factor. After delivering its payload of food, clothing, propellant fuel, etc., the spacecraft was unloaded and over the next few months refilled with unwanted materials and waste. Shortly after this, the *Soyuz TMA-04* crew returned to Earth on 17 September, landing in Kazakhstan.

It wasn't until 9 February 2013 before *Progress M-16M* was undocked and placed in a decaying orbit to be burnt up on re-entry. In the meantime a new crew had arrived on 25 October 2012. *Soyuz TMA-06M* was launched on a Soyuz-FG rocket on 23 October 2012 to deliver the *Expedition 33* crew of Oleg Novitskiy (R), Evgeny Terelkin (R) and Kevin Fod (A) to the ISS. The spacecraft docked in the Poisk module, which is attached to the Zenith port of the Zvezda module, using the Kurs automated docking rendezvous system. After carrying out all the usual safety checks, the relief crew entered the ISS to be greeted by a smiling resident crew. One month later, on 18 November, the *Soyuz TMA-5M* crew handed over command of the space station, undocked from the ISS and landed in Kazakhstan the following day. In the meantime another cargo spacecraft had arrived at the space station on 31 October, *Progress M-17M*. This was another of the rapid approach missions with the spacecraft docking in the aft port of the Zvezda module the same day. After being unloaded the cargo spacecraft was reloaded over the next few months with waste materials and rubbish, but remained docked to the space station until 15 April 2013 when it was undocked and placed in a decaying orbit to burn up on 21 April over the Pacific Ocean. Whilst the cargo spacecraft remained docked, a new crew had arrived in *Soyuz TMA-07M*, which had been launched on 19 December 2012 on a Soyuz-FG rocket and delivered the *Expedition 34* crew of Roman Romanenko (R), Chris Hadfield (Can) and Thomas Marshburn (A) to the ISS. After a two-day flight, as they approached the ISS, the Kurs-A Navigation System on the space station and the Kurs-P System on the Soyuz *TMA-07M* were activated and automatically docked the spacecraft with the nadir port of the Rassvet module. After carrying out the usual safety checks, the new crew cracked the hatch to be greeted by the resident crew. The two crews then settled down to their respective tasks, many of which involved physiological and cardiovascular tests.

The next to visit the ISS was the *Progress M-18M* cargo spacecraft that had been launched on a Soyuz-U rocket, arriving on 12 February 2013 with much needed supplies for the crew of the ISS. After docking in the nadir port of the Pirs module, the cargo spacecraft was unloaded, and over the next few months, refilled with waste and unwanted materials. One month later, on 15 March 2013, the resident crew handed over command of the space station to Roman Romanenko, climbed into their spacecraft *Soyuz TMA-06* and undocked from the ISS. They landed the following day near Arkaylk, Kazakhstan. Two weeks later *Soyuz TMA-08M* was launched on a Soyuz-FG rocket on 28 March 2013, to deliver the *Expedition 35* crew of Pavel Vinogradov (R), Alexsander Misurkin (R) and Christopher Cassidy (A) to the ISS. This was one of the rapid missions to the ISS and the spacecraft docked in the Zenith port of the Poisk module the following day. After completing all the safety checks, the hatch was opened and the relief crew were welcomed by the resident crew. They then all participated in the now traditional welcoming ceremony that involved family members and officials back at Mission Control. The arrival of *Progress*

M-19M on 26 April brought fresh supplies and more equipment to the ISS. Launched on the 24 April on a Soyuz-U rocket, the cargo spacecraft docked in the aft port of the Zvezda module. With the docking successful and all safety checks completed, the crew entered the cargo spacecraft. The two crews then removed 106lbs (48 kilograms) of oxygen, 805lbs (365 kilograms) of propellant, 900lbs (410 kilograms) of water and some scientific equipment as well as food and spare parts. Two months later, on 13 May, the *Expedition 34* crew undocked *Soyuz TMA-07* from the space station, and after handing over command, said their farewells and returned to Earth the following day, landing near Dzhezkazgan, Kazakhstan. Many of the experiments that were carried out on the ISS, aside from the medical and scientific ones, were either Department of Defense (DoD) or were secret.

The next crew to arrive was *Expedition 36* on *Soyuz TMA-09M*. Launched on a Soyuz-FG carrier rocket on 28 May 2013, it delivered the crew of Fyodor Yurchikhin (R), Luca Parmitano (It) and Karen Nyberg (A) to the ISS. The spacecraft docked in the nadir port of the Rassvet module just six hours after being launched, minimizing the time the crew had to spend in the cramped conditions of the Soyuz spacecraft. After carrying out the usual safety checks, the crew cracked the hatch and entered the space station to be greeted by the resident crew. After the greeting ceremony the crew got down to work with their experiments and of course the normal housekeeping duties. This included filling the now empty *Progress M-19M* cargo spacecraft with waste and unwanted materials, but it was left there until the 11 June before it was undocked. The cargo spacecraft then went into an independent orbit and began a one week free flight to be part of a Radar–Progress Experiment. The object of the experiment was to measure the density, reflection and size of the ionosphere environment around the spacecraft due to engine burns. On the 19 June with all the experiments completed, the Progress spacecraft went into a decaying orbit and its remains plunged into the Pacific Ocean after being burnt up on re-entry. *Progress M-18M* had remained docked until 26 July when it was undocked and placed in a decaying orbit before burning up over the Pacific Ocean the following day. One month later another cargo spacecraft arrived, *Progress M-20M*. Launched on 27 July 2013 on a Soyuz-U companion rocket, it docked six hours later in the nadir port of the Pirs module. This was one of the rapid flights to the space station which were now becoming the norm. After being unloaded, the spacecraft was filled over the next seven months, with the usual unwanted rubbish and materials before the hatch was closed. On 11 September the *Expedition 35* crew of Vinogradov, Misurkin and Cassidy handed over command to Fyodor Yurchikhin, got into their Soyuz spacecraft and undocked from the space station. They landed near the town of Dzhezkazgan, Kazakhstan five hours later. The weather in the area at the time was very bad and the three crew members were quickly removed from their space capsule and taken to a medical tent where they were checked over by medical teams. As with all the crews returning from the space station, they were put through a series of exercises to assess their physiological state. Two weeks later *Soyuz TMA-10M* was launched on 25 September 2013, to deliver the *Expedition 37* crew of Oleg Kotov (R), Sergey Ryazansky (R) and Michael Hopkins (A) to the ISS. Launched from Baikonur on a Soyuz-FG rocket, the spacecraft docked in the Zenith port of the Poisk module just six hours later. After the usual safety checks, the crew opened the hatch and entered the space station to be met by a smiling resident crew. The six members of the space station quickly got down to their allotted tasks, which included the necessary housekeeping duties. The *Soyuz TMA-09* spacecraft was moved on 1 November from the Rassit module to the aft port of the Zvezda module to make way for the arrival of *Soyuz TMA-11M*. Launched on 7 November 2013 to deliver the *Expedition 38* relief crew of Mikhail Tyurin (R), Richard Masstraccio (A) and Koichi Wakata (Jpn) to the ISS. For the very first time, *Soyuz TMA-11M* carried the Olympic flame for the 2014 Winter Olympics into space. The Olympic torch returned to Earth five days later aboard *Soyuz TMA-09M*, which undocked from the ISS on 10 November with crew members Fyodor Yurchikhin, Luca Parmitano and Karen Nyberg aboard, and after a five hour flight touched down near Dzhezkazgan, Kazakhstan. Just over two weeks later, on 25 November 2013, the next cargo spacecraft, *Progress M-21M*, was launched from Baikonur on a Soyuz-U rocket. This was another of the rapid missions to the ISS and the cargo spacecraft docked in the nadir port of the Zvezda module just six hours later. After unloading all the supplies, the spacecraft was refilled with unwanted materials and waste and the hatch closed. It was to remain docked with the ISS until 9 June 2014. Two days before the arrival of the next cargo spacecraft, *Progress M-20M* undocked on the 3 February 2014 and was allowed to go into a decaying orbit. The next spacecraft to visit the space station was *Progress M-22M*, which arrived two days later. Launched from Baikonur on a Soyuz-U rocket, the cargo spacecraft reached the ISS in just over six hours and docked in the nadir port of the Pirs module. After being unloaded and refilled with the usual unwanted waste, the spacecraft was undocked and placed in a decaying orbit on 7 April to burn up over the Pacific Ocean. The *Progress M-21M* however remained docked, then on 13 March its engines were fired for ten minutes and used to raise the orbit of the space station by a distance of two kilometres. There had earlier been a problem with the Kurs-NA docking system, so it was decided to test the system again. The *Progress M-21M* cargo spacecraft was undocked from the Zvezda module on 23 April to test the Kurs-NA rendezvous docking system and successfully re-docked on 25 April. It was finally undocked on 9 June and placed in a decaying orbit to be burnt up on re-entry over the Pacific Ocean.

The next manned spacecraft to visit the ISS was *Soyuz TMA-12M*, launched on a Soyuz FG rocket on 25 March 2014 to deliver the *Expedition 39* relief crew of Aleksandr Skvortsov (R), Oleg Artemyev (R) and Steven Swanson (A). The spacecraft docked two days later in the zenith port of the Poisk module. The spacecraft had been scheduled to dock with the space station on 26 March but was unable to carry out a third mid-course correction because of a problem with the attitude control system. The following day, using the Kurs-NA automatic rendezvous docking system, *Soyuz TMA-12M* successfully docked with the ISS. After completing the usual safety checks, the crew opened the hatch and entered the space station to be greeted by the resident crew. Like all the previous crews, aside from the medical and physiological experiments, little is known about the remainder of each country's experiments. Two weeks after the arrival of

Soyuz *TMA-12M*, on 9 April 2014 the *Progress M-23M* cargo spacecraft was launched on a Soyuz-U rocket. This was another one of the six-hour flights to the space station and it successfully docked in the nadir port of the Pirs module right on schedule. After being unloaded the spacecraft was refilled with waste and unwanted materials. The hatch was closed on 21 July and it undocked from the ISS. It then participated in the Radar–Progress Experiment before deorbiting on 28 July to be burnt up over the Pacific Ocean on re-entry. In the meantime *Soyuz TMA-13M* had been launched on a Soyuz-FG rocket on 28 May 2014 to deliver the *Expedition 40* crew of Maksim Surayev (R), Alexander Gerst (Ger) and Gregory Wiseman (A). This was a rapid flight of just under six hours to the ISS and the spacecraft docked successfully right on schedule in the Rassvet module. After the usual safety checks had been carried out, the hatch was cracked two hours later and the relief crew were welcomed by the resident crew. With formalities over, both crews settled down to carry out their tasks and experiments and their share of housekeeping duties. The latter was a very important part of living in such a confined space with so much equipment lying about. The cargo spacecraft *Progress M-24M*, which had been launched from Baikonur on 23 July on a Soyuz-U rocket, docked in the Pirs module just six hours after being launched. It carried the usual fresh supplies and some maintenance equipment to a total of 2,322 kilograms. It brought welcome fresh food supplies, fresh clothing and personal letters and items for the crews. After being unloaded it was refilled with rubbish and other waste materials. The hatch was closed on 27 October and then undocked, leaving the cargo spacecraft to drift in a slow decaying orbit, but it wasn't until the 19 November before it finally deorbited and was burnt up on re-entry over the Pacific Ocean.

The *Expedition 41* relief crew arrived on the ISS aboard *Soyuz TMA-14M*, which was launched on 25 September 2014 on a Soyuz-FG rocket from Baikonur. This was another of the six-hour rapid flights to the space station that was designed to relieve the stress on the crew of Aleksandr Samokutyayev (R), Yelena Serova (R) and Barry Wilmore (A). After carrying out four orbits of the Earth, the spacecraft docked in the Zenith port of the Pirs module right on schedule using the Kurs automatic docking system. With the usual safety checks completed, the crew entered the space station to be greeted by the resident crew. After the welcoming ceremony the crews got down to their allotted tasks. One month later, on 29 October 2014, *Progress M-25M* was launched on a Soyuz-U rocket. This was another of the six-hour rapid flights and the cargo spacecraft docked successfully with the Pirs module right on schedule. After being unloaded with the usual supplies of fresh food, water, fuel, experimental hardware and personal items, the cargo spacecraft was refilled over the next five months with waste and other unwanted materials. The spacecraft remained docked until 25 April 2015.

In the meantime *Soyuz TM-15M* was launched on 23 November 2014 to deliver the *Expedition 42* crew of Anton Shkapierov (R), Samantha Christoforetti (It/ESA) and Terry Virts (A) to the ISS. This was another six-hour rapid flight to the ISS and the spacecraft docked right on schedule in the nadir port of the Rassvet module. After the usual safety checks had been carried out the crew emerged from their spacecraft to be greeted by the resident crew. Samantha Christoforetti presented the resident crew with three Lego figures of herself, Anton Shkapierov and Terry Virts, that had been specially commissioned by the ESA (European Space Agency). With the formalities over, the crews got on with their tasks, which included the general housekeeping duties that all crews faced and the cardiovascular and physiology experiments on each other. On 17 February 2015, *Progress M-26M* was launched on a Soyuz-U rocket with supplies for the crews aboard the ISS. The cargo spacecraft arrived in less than five hours after being launched and docked in the aft port of the Zvezda module. It carried a total of 2,370 kilograms of cargo made up of food, water, propellant, personal items and other equipment. After being unloaded the now empty spacecraft was filled over the next seven months with unwanted materials and waste. During this period the next manned space flight to the ISS arrived on 27 March 2015, when *Soyuz TMA-16M* was launched on a Soyuz-FG rocket, to deliver the *Expedition 43* crew of Gennady Pedalka (R), Mikhail Korniyenko (R) and Scott Kelly (A). This was to be the last space mission for all three cosmonaut/astronauts. After a rapid flight to the space station of just over six hours, the spacecraft docked in the zenith port of the Poisk module right on schedule. With the usual safety checks completed the relief crew entered the space station two hours later to be greeted by the resident crew. With the formalities over the two sets of crews got down to their tasks.

The launch of *Progress M-27M* on 27 March 2015 went disastrously wrong just minutes after lift-off, when a malfunction occurred close to the upper stage burn shortly before separation, leaving the cargo spacecraft spinning and dangerously out of control. It was destroyed by mission control. In May, whilst still docked, the *Progress M-26M* was used to re-boost the station. The first attempt was automatically aborted just one second into the burn because of one of the eight thrusters had failed. A second attempt using just seven of the thrusters was successful. The cargo spacecraft remained docked to the space station until 14 August when it was undocked and placed in a decaying orbit to burn up over the Pacific Ocean. On 10 June, during preparation for the return of the *Soyuz TMA-15M* crew, they accidentally fired its engines one day before undocking. The one minute burn slightly shifted the station's position, but did not prevent the undocking of the spacecraft the following day with its returning crew of Anton Shkapierov, Samantha Christoforetti and Terry Virts to Earth. The spacecraft landed near Arkalyk, Kazakhstan on 11 June.

The launch of *Soyuz TMA-17M* had been planned for 26 March, but since the loss of the *Progress M-27M*, it had been delayed because of suspected space debris. It was finally launched from Baikonur on a Soyuz-FG rocket on 22 July 2015. The crew of Oleg Konenko (R), Kimiya Yui (Jpn) and Kjell Lindgren (A) were the *Expedition 44* relief crew. The spacecraft docked successfully in the nadir port of the Rassvet module after a six-hour flight. After the usual safety checks, the hatch was opened and the oncoming crew were greeted by the resident crew. Formalities over, the crews got down to the scheduled tasks and the everyday housekeeping duties. On the 4 July *Progress M-28M* arrived after being launched on a Soyuz-U rocket six hours earlier in the early hours of 3 July 2015, docking in the nadir port of the Pirs module. Amongst the desperately needed fresh supplies and normal cargo being carried to the ISS, there was experimental hardware for the crew of *Expedition 44*. After being unloaded and refilled with unwanted materials and waste, the hatch was closed. It wasn't until 19 December before the cargo spacecraft undocked and was placed in a decaying orbit to be burnt up on re-entry. On 28 August the *Soyuz TMA-16M* spacecraft was moved from the Poisk module to the aft port of the Zvezda module to make way for the arrival of the next spacecraft – *Soyuz TMA-18M*. Launched on a Soyuz-FG rocket on 2 September to deliver the *Expedition 45* crew of Sergey Volkov (R), Andreas Mogensen (ESA), Aldyn Aimbetov (Kaz) to the ISS. The third seat had originally been booked for a space tourist, the British singer Sarah Brightman, but she withdrew from training and dropped out. Her place was then offered to Japanese entrepreneur Satoshi Takamatsu, who had been training as Sarah Brightman's back-up. He then withdrew from the program as the art project he had planned would have not been ready in time for the launch. Russian businessman Fjlaret Galchev was then offered the seat, but turned it down saying that he did not have the time for the compulsory training that was required for the mission. It was then given to Kazakh cosmonaut Aidyn Aimbetov. After docking with the space station in the Poisk module, and after going through the safety procedures, the relief crew cracked the hatch to be welcomed by the resident crew. With the formalities over, the two crews settled down to their tasks, which included the ongoing household duties and maintenance. *TMA-16M* was to stay docked until 11 September when it returned to Earth with Gennady Padalka, Andreas Morgensen and Aidyn Aimbertov, landing near Arkalyk, Kazakhstan just three hours after undocking from the space station.

Another of the unmanned Progress cargo spacecraft was launched on 1 October 2015, *Progress M-29M*. The cargo spacecraft lifted off the launch pad at Baikonur on a Soyuz-U rocket and after a six hour flight docked in the aft port of the Zvezda module. Amongst the 2,369 kilograms of fresh supplies, spare parts, oxygen, fuel propellant, etc., were a number of pieces of experimental hardware for the crew to evaluate. After being unloaded, the spacecraft was refilled with waste and unwanted materials over the next six months before the hatch was closed. It was to remain attached to the space station before undocking on 30 March 2016 and going into a decaying orbit to be burnt up on re-entry. In the meantime *Soyuz TMA-19M* had been launched on 15 December 2015 on a Soyuz-FG rocket to deliver the *Expedition 46* crew of Yuri Malenchenko (R), Timothy Kopra (A) and Timothy Peake (UK) to the ISS. Tim Peake's trip was the first for a British cosmonaut since Helen Sharman's visit to the *Mir* space station aboard *Soyuz TM-12* in May 1991. As their spacecraft prepared to dock in the nadir port of the Rassvet module, the Kurs docking navigational system failed. Yuri Malenchenko immediately took over control and manually docked the spacecraft. Time spent in the simulator back on Earth once again proved to have been well spent. Safely docked, the crew went through the usual safety checks before cracking the hatch and entering the space station. One week later *Progress MS-01* was launched on a Soyuz-2.1a rocket on 21 December 2015. The cargo spacecraft docked two days later on 23 December in the nadir port of the Pirs module. The *Progress MS-01* was an updated version of the Progress M model, which in essence was a complete upgrade of the communication and navigation systems. The changes did not alter the external appearance of the Soyuz spacecraft, except for the deployed antenna. These updates were also incorporated into the manned Soyuz spacecraft. After being unloaded, the crew filled the cargo spacecraft with unwanted rubbish over the next six months before closing the hatch. On 15 January 2016, Timothy Kopra and Timothy Peake carried out an EVA to replace a failed voltage regulator on one of the modules. This was the first official spacewalk to be carried out by a British cosmonaut who was wearing the Union Jack on his EMU. The EVA had to be cut short after Timothy Kopra reported water in his helmet. Kopra was wearing the same EVA suit that had been worn by Luca Parmitano of Expedition 36, when his helmet became flooded during his EVA. There was one light moment when Scott Kelly chased Tim Peake through the space station whilst wearing a gorilla suit.

On 1 March 2016, Scott Kelly transferred command of the ISS to Timothy Kopra, then together with Mikhail Kornienko and Sergei Volkov climbed into their *Soyuz TM-18M* spacecraft, undocked and returned to Earth. Their spacecraft landed near the remote town of Dzhezkazgan, Kazahkstan three hours later. Scott Kelly returned to Houston, Texas the following day.

Just over two weeks later, on 18 March 2016, *Soyuz TMA-20M* was launched on a Soyuz-FG rocket, to deliver the *Expedition 47* crew of Aleksey Ovchinin (R), Oleg Skripochka (R) and Jeffrey Williams (A) to the ISS. This was the final flight of the Soyuz TMA series of spacecraft, having been superseded by the Soyuz MS model. After a six hour flight their spacecraft docked in the Zenith port of module. After the usual safety checks had been completed, the ongoing crew cracked their hatch and entered the space station to be greeted by the resident crew. It was to be a further two weeks before the next visitor to the space station arrived and that was the unmanned cargo spacecraft *Progress MS-02*, which was launched on a Soyuz 2.1a rocket on 31 March 2016. After a two-day flight the cargo spacecraft docked in the aft port of the Zvezda module. The upgraded communications on the Progress cargo spacecraft enabled ground control to communicate with the spacecraft via a geosynchronous relay satellite that was in orbit at the time. This was the first time that a Russian spacecraft had such a capability. Amongst the other modifications was an upgraded Kurs-NA rendezvous automatic docking system, new digital television system, an enhanced meteoroid shield, LED lighting system and new docking port equipment. The cargo spacecraft was carrying 2,425 kilograms of cargo and supplies for the space station, which included food, fuel, oxygen, water, spare parts, experimental hardware and an amateur satellite, Tomsk-TPU, that had been built by the Tomsk Polytechnic University of Tomsk, Siberia. After being unloaded, the cargo spacecraft was refilled over the next seven months with unwanted materials and waste. Earlier, on 18 June, the crew of *Soyuz TMA-19M*, Timothy Peake, Timothy Kopta and Yuri Malenchenko, said their farewells to the other crew members of the space station, climbed into their spacecraft and headed back to Earth. They safely landed three hours later near Karaganda, Kazakhstan. Two weeks later the *Progress MS-01* cargo spacecraft, which had remained docked with the ISS until 2 July 2016, was undocked and placed into a decaying orbit to be burnt up on re-entry. Five days later *Soyuz MS-01* was launched on a Soyuz-FG rocket on 7 July 2016, to deliver the *Expedition 48* crew of Anatoli Ivanishin (R), Takyua Onishi (Jpn) and Kathleen Rubins (A), to the ISS. This was the first flight of the new manned Soyuz MS spacecraft. The launch had been scheduled for 6 June, but teething problems with the new spacecraft's control system that could have affected the docking to the ISS, caused it to be re-scheduled. Their flight to the space station was not one of the rapid ones and the spacecraft docked in the nadir port of the Rassvet module on 9 July. After going through the usual safety procedures, the relief crew cracked the hatch and entered the space station to be greeted by the resident crew. With the formalities over, both crews settled down to their allotted tasks, which included the essential housekeeping duties. More supplies and equipment arrived on *Progress MS-03*, which had been launched on a Soyuz-U rocket on 16 July 2016. This was another of the 'slow' flights to the space station and the cargo spacecraft docked in the nadir port of the Pirs module two days later after carrying out thirty-four orbits of the Earth. It was a great relief to the crew of the ISS as they were beginning to run low on essentials. The launch had originally been scheduled for 30 April, but was postponed as a result of an overall reshuffle of the flight manifest for the International Space Station. After being unloaded, the cargo spacecraft was refilled with waste and unwanted materials. It was to remain docked with the ISS until 31 January 2017 when it was released and sent into a decaying orbit to be burnt up over the Pacific Ocean.

With all their experiments completed, the crew of *Soyuz TMA-20M* handed over command of the space station to the crew of *Soyuz MS-01* and on 7 September said their farewells, climbed into their spacecraft and returned to Earth, landing in Kazakhstan.

Progress MS-02, which had remained connected to the space station since 2 April, was undocked on 14 October and allowed to go into a decaying orbit before burning up on re-entry over the Pacific Ocean. One week later came the next arrival, *Soyuz MS-02*, that had been launched on a Soyuz-FG rocket on 19 October 2016 to deliver the *Expedition 49* crew of Sergey Ryzhikov (R), Andrei Borisenko (R) and Shane Kimbrough (A). The launch had been originally scheduled for 23 September 2016, but during final tests a mysterious short circuit appeared. The mission was immediately postponed and an investigation started. The problem was discovered to be a bent wire behind the cosmonauts' seats in the descent module. Engineers repaired the fault, despite ignoring safety rules that prohibited working on the spacecraft whilst it was loaded with toxic propellants and pressurized gases. After a two day flight the spacecraft docked in the Zenith port of the Pirs module using the Kurs automated docking system. During the flight, the crew carried out tests on a number of the upgraded systems. With the spacecraft safely docked and the safety checks completed, the hatch was opened and the relief crew entered the space station. With the greeting formalities over, the crews got down to the everyday business of running a space station. On 30 October 2016, with all their experiments completed, the crew of *Soyuz MS-01* handed over command of the space station, said their farewells and undocked and returned to Earth, landing on the steppes of Kazakhstan. One month later, on 17 November 2016, *Soyuz MS-03* was launched on a Soyuz-FG rocket to deliver the *Expedition 50* crew of Oleg Novitsky (R), Thomas Pesqueto (Fr/ESA) and Peggy Whitson (A) to the ISS. At the age of 56, Peggy Whitson became the oldest woman to fly in space. After docking in the nadir port of the Rassvet module, and carrying out the required safety checks, the relief crew cracked the hatch and entered the space station. After the usual greetings and formalities had been carried out, the crews settled in to their roles on the space station.

The next spacecraft scheduled to arrive at the ISS was the cargo spacecraft *Progress MS-04*. Launched 1 December 2016 on a Soyuz-U rocket, the cargo spacecraft lifted off the launch pad at Baikonur right on schedule carrying 2,540 kilograms of supplies. Two minutes later at the point of the second burn, the third stage separated from the second stage six minutes early. A report from the town of Tuva said of a violent explosion in the air with the debris crashing down in a mountainous region approximately 3,500 kilometres downrange from Baikonur. The initial investigation found that the Progress spacecraft had separated from the third stage six minutes and thirty seconds into the launch, but the third stage telemetry had failed to respond. The reason for the premature separation was unclear, but what was also unclear was why the Blok 1 computer system didn't issue a manual shut-off command, which it had the capability to do in the event of a malfunction. After a detailed investigation into the demise of *Progress MS-04*, the government ordered the recall of all Proton-M second and third stage engines produced by the Voronezh Mechanical Plant. They also ordered the disassembly of three completed Proton rockets and a three and half month suspension of flights. A further investigation found that engine parts that were supposed to be made of precious metals, were in fact being substituted with cheaper alternatives that were unable to withstand high temperatures. In addition to this they found that the production and certification documents had been falsified. The Progress rocket had been insured for 2,135 billion rubles, so it would be interesting to know who ended up paying for the disaster and if there were any repercussions concerning the company itself. Such was the desperation to get fresh supplies to the ISS, that on 9 December 2016, the Japanese Aerospace Exploration Agency (JAZA) launched an unmanned cargo spacecraft, *HTV-6* (*Kounotori-6*), to the ISS loaded with more than 4.5 tons of food, water, spare parts and experiments for the six crew members aboard the space station. Also included in the cargo were six lithium-ion batteries to replace the nickel-hydrogen batteries that were currently being used to store electrical energy generated by the space station's solar array. They were later installed during a series of spacewalks. As the Japanese spacecraft approached the ISS, it was captured by the Canadarm 2 robotic arm and installed in the Harmony module. After being unloaded and refilled with waste and unwanted materials over the next two months, it remained docked until the 3 February. *Progress MS-03*, which had been docked since the 16 July 2016, was undocked on 31 January and placed in a decaying orbit as it entered the Earth's atmosphere over the Pacific Ocean. Three days later the Japanese cargo spacecraft HTV-6, was undocked using the Canadarm. After being moved to a safe distance from the ISS, the spacecraft started to demonstrate the Kounotori Integrated Tether Experiment using an electrodynamic tether. The object of the exercise was to capture space debris, which was becoming a major cause for concern. It wasn't successful, so the cargo spacecraft was allowed to go into a decaying orbit and on 3 February 2017 it was burnt up over the Pacific Ocean as it re-entered the Earth's atmosphere.

It was to be almost three months before another Russian cargo spacecraft arrived at the ISS, when *Progress MS-05* arrived on 24 February 2017. Launched from Baikonur on a Soyuz-U rocket the unmanned cargo spacecraft docked in the nadir port of the Pirs module after a two-day flight, right on schedule. Alongside the usual cargo of food, water and spares was an Orlon-MKS spacesuit to replace the one lost when *Progress MS-04* exploded and was lost. After being unloaded, the cargo spacecraft was refilled with unwanted materials and waste. It remained docked to the space station for the next five months.

On 9 April 2017, *Expedition 50* commander Shane Kimbrough handed over command of the space station to Peggy Whitson. The following day, together with Sergei Ryzhikov and Andrei Borisenko, he climbed into the *Soyuz MS-02* spacecraft and undocked from the space station to head for Earth. There was a minor problem during the descent, when the capsule was partially depressurized when the main parachute was deployed, but the spacecraft landed safely near Dzhezkazgan, Kazakhstan.

The arrival of another cargo spacecraft, *Cygnus OA-7*, brought fresh supplies. Launched from Cape Canaveral on the 18 April 2017, it reached the space station on 22 April and once again the Canadárm-2 was used to grasp the spacecraft and dock it in the nadir port of the Unity module. The cargo spacecraft carried a total of 3,459kg of pressurized and unpressurized material and supplies for the ISS crew. In addition it also carried four CubeSats that would be released when *Cygnus-7* undocked. This spacecraft was launched by NASA under a commercial agreement to supply the ISS.

The next crew to visit the ISS were the *Expedition 52* crew of Fyudor Yurchikhin (R) and Jack Fischer (A), who were launched aboard *Soyuz MS-04* on a Soyuz-FG rocket on 20 April 2017. This was another of the six-hour flights and the spacecraft docked in the zenith port of the Poisk module without incident. After the usual safety checks, the two members of the relief crew cracked the hatch and entered the space station to be greeted by the resident crew. With the formalities over, the five members of the crew settled down to their tasks. Peggy Whitson carried out a series of physiological tests concerned with a new type of lighting that had been installed and the effects, or non-effects, it had on her fellow crew members. This was amongst the day-to-day housekeeping duties everyone was involved with. Two members of the *Expedition 50* crew, Oleg Novitsky (R) and Thomas Pesquet (Fr/ESA), said their farewells to the other members of the space station on 2 June 2917, climbed into *Soyuz MS-03*, undocked and returned to Earth three hours later, landing in Kazakhstan, whilst the third member of the original crew, Peggy Whitson, remained on the ISS.

The cargo spacecraft *Cygnus OA-7*, after being filled with unwanted materials and general rubbish, was undocked on 11 June but prior to being placed into a decaying orbit, it released the four CubeSats and then entered the Earth's atmosphere to be burnt up on re-entry. Two weeks later the cargo spacecraft *Progress MS-06*, was launched on 14 June on a Soyuz-2.1a rocket and after a two-day flight docked in the nadir port of the Zvezda module. This was another one of the models that carried an external compartment containing four launch containers that enabled satellites to be deployed. It was supposed to dock in the Pirs module and then remove the whole module to make way for its replacement, the Nauka module, but because of repeated delays by the manufacturers of the Nauka, the move was delayed to the launch of *Progress MS-09*. After being unloaded *Progress MS-06* was refilled over the next few months with waste and rubbish. *Progress MS-05*, which had been docked with the ISS since the 24 February, was undocked on 20 July and placed in a decaying orbit to burn up on re-entry over the Pacific Ocean.

The next crew to arrive on the International Space Station was *Expedition 53* in *Soyuz MS-05*. Launched from Baikonur on a Soyuz-FG rocket on 28 July 2017, the flight took just over six hours and docked in the nadir port of the Rassvet module. After carrying out the usual safety checks, the crew of Sergei Ryanansky (R), Paolo Nespoli (It) and Randy Bresnik (A), cracked the hatch and entered the space station. After the formalities had been completed, the two crews got down to their tasks and experiments, most of the latter being concerned with the effects that living in a weightless environment had on the human body. They also all had to share the day-to-day duties of running the space station itself. With all their tasks completed the resident crew of Fyodor Yurchikhin, Jack Fischer and Peggy Whitson handed over command of the space station and prepared to return to Earth. Peggy Whitson had spent 289 days in space, which was the longest single space flight by a woman. On 2 September 2017, the three cosmonaut/astronauts climbed into *Soyuz MS-04* and undocked from the space station. They landed near Dzhezkazgan, Kazakhstan three hours later. Nine days later on 12 September *Soyuz MS-06* was launched on a Soyuz-FG rocket to deliver the *Expedition 53* crew of Aleksandr Mizurkin (R), Mark Vande Hei (A) and Joseph Acaba (A) to the ISS. After a six-hour flight, the spacecraft docked in the zenith port on the Poisk module. With the usual safety checks completed, the relief crew cracked the hatch to be greeted by the resident crew. With the formalities over the six-man crew got down to their allotted tasks and experiments.

The arrival of *Progress MS-07* brought fresh supplies and materials to the ISS. The cargo spacecraft was launched on 14 October 2017 on a Soyuz-2.1a rocket and, after a two-day flight, docked in the nadir port of the Pirs module. This cargo spacecraft, like its predecessors MS-03, 04 05, and 06, carried an external compartment that allowed it to launch satellites. With the cargo craft securely docked, it was then unloaded and then refilled with waste and unwanted materials. On 20 October, the commander of the space station, Randolph Breznik, and flight engineer Joseph Acaba, undertook a spacewalk to carry out a number of maintenance tasks that included the replacement of a fuse on the robotic Canadarm-2, replacing an external camera and light fitting and removing thermal insulation from two spare units in preparation for future relocation. It was to remain docked until 26 April 2018.

Above: Rear view shot of *Soyuz MS-06* on board its specially constructed railway equipment, being taken to the launch site. (Roscosmos)

Below: *Soyuz MS-06* being transported to the launch site. (Roscosmos)

Soyuz MS-06 about to be launched from Baikonur. (Roscosmos)

On the 14 December 2017, Sergei Ryanansky (R), Paolo Nespoli (It) and Randy Bresnik (A), passed command of the ISS to the relief crew and, after saying their farewells, clambered into their Soyuz MS-05 spacecraft and returned to Earth. The spacecraft stopped some distance from the space station for twenty-five minutes to enable Pablo Nespoli to take a series of photographs of the ISS. Their spacecraft touched down in Kazkhstan just over three hours later. The launch of *Soyuz MS-07* from Baikonur on a Soyuz-FG rocket on 17 December was to deliver the *Expedition 54* crew of Anton Shkaplerov (R), Scott Tingle (A) and Norishie Kanai (Jpn), to the ISS. After a two-day flight, the Soyuz spacecraft docked in the nadir port of the Rassvet module. After the usual, but obligatory safety checks, the hatch was cracked and the relief crew were welcomed by the resident crew. With the formalities completed, the six-man crew settled down to carry out a variety of physiological experiments, as well as the normal maintenance duties. *Progress MS-06* 'which had remained docked since 16 June' was undocked on 28 December and sent into a decaying orbit before burning up over the Pacific Ocean.

The first of the EVAs took place on 2 February 2018 when Aleksandr Mizurkin, from the *Soyuz MS-06* crew, and Anton Shkaplerov, began a spacewalk to dismantle and discard a ShA-317A-II radio receiver on the Lira high-gain antenna which was mounted in the instrument section of the Zvezda module. Once removed, they discarded the sixty pound piece of equipment into space at an angle that prevented it from hitting the space station and left it to de-orbit and burn up on re-entry. They then installed a new wide-band communication system on the Lira antenna which gave the Russian ground control, via the Luch relay satellites, almost round-the-clock monitoring of transmissions from the Russian segment of the International Space Station. Two weeks later the unmanned cargo craft, *Progress MS-08*, was launched on 14 February on a Soyuz-2.1a rocket. After a two-day flight the spacecraft docked in the aft port of the Zvezda module. Amongst the 2,494 kilograms of food, water, fuel and oxygen were two nano-satellites, Tanyusha YuZGU-3 and Tanyusha YuZGU-4, which had been developed jointly between RKS Energia and students of the South-Western State University of Kursk. Both satellites were later deployed during spacewalks. After the cargo had been unloaded, they were refilled with waste and unwanted materials. On 28 March, *Progress MS-07* was undocked from the ISS and before going into a decaying orbit, successfully launched the satellite she had been carrying in her special compartment. With the deployment completed the cargo spacecraft descended into the Earth's atmosphere amd was burnt up on re-entry over the Pacific Ocean.

With the relief crew scheduled to arrive in March, the *Soyuz MS-06* crew made preparations to return to Earth. On 27 February 2018, the three crew members, Aleksandr Mizurkin, Mark Vande Hei and Joseph Acaba, handed over command of the space station to Anton Shkaplerov, Scott Tingle and Norishie Kanai, and said their farewells. They then climbed into their spacecraft, undocked and three hours later landed near Dzhezkazgan, Kazakhstan. Three weeks later, the relief crew of *Expedition 55* were launched to the ISS in *Soyuz MS-08* on a Soyuz-FG rocket on

21 March 2018. After a two-day flight, the spacecraft docked in the zenith port of the module. With all the safety checks completed, the crew of Oleg Artemyev (R), Andrew Feustal (A) and Richard Arnold (A), cracked the hatch and entered the space station to be greeted by the resident crew. After going through all the formalities, both crews settled down to their work, which included the monotonous, but necessary housekeeping and maintenance duties.

It was to be two months before the next visitors arrived at the space station, in the shape of *Soyuz MS-09*. Launched on a Soyuz-FG rocket on 6 June 2018 carrying the *Expedition 56* crew of Sergei Prokopyev (R), Alexander Gerst (Ger) and Serena Auñón-Chancellor (A), the spacecraft docked in the nadir port of the Rassvet module. With the safety checks completed, the relief crew cracked the hatch and entered the space station to be greeted in the customary manner by the resident crew. With the formalities over the six crew members settled into their routine of working. On 3 June the crew of *Soyuz MS-07* handed over command of the space station to the *Expedition 56* crew and with that Anton Shkaplerov, Scott Tingle and Norishie Kanai said their farewells, climbed into their spacecraft and undocked from the space station. Three hours later they landed safely in Dzhezkazgan, Kazakhstan.

A welcome distraction in the shape of *Progress MS-09* arrived on 9 July 2018. Launched on a Soyuz-2.1a rocket, the cargo spacecraft docked in the nadir port of the Pirs module in just three hours and forty minutes – making it the fastest rendezvous with the International Space Station ever attempted. Incorporated into the latest model was an upgraded version of the Kurs-NA which increased reliability and safety during docking. Communication improvements had also been made to upgrade the Luch-5 data relay satellites. With the cargo spacecraft docked, the crew started to unload the 2,450 kilograms of cargo consisting of food, water, personal items and fuel. The now empty cargo spacecraft was slowly refilled with waste and unwanted items over the next few months whilst the crew settled down to their tasks. The cargo spacecraft had been scheduled to remove the Pirs module, but once again delays by the manufacturers of its replacement caused the removal to put on hold.

CHAPTER NINE

The Mystery of the Hole

Then on 29 August an incident occurred when suddenly an air leak was discovered coming from a two-millimetre-wide hole in the orbital module. One of the Russian crew members patched it up temporarily with tape, but it was secured and sealed permanently later using an epoxy resin sealant. On further examination it was declared that the hole had been made by a drill, but was unclear as to whether or not it was an accident or deliberate. In ground control, rumours amongst the officials were rife as speculation hinted that it could possibly have been sabotage and that one of the American crew members was responsible. An EVA by Konenenko and Prokopyev went to look at the possible damage from the outside, by cutting the thermal blankets and pulling apart the insulation around the area to examine the external hull. The hole was in the orbital module that was jettisoned before re-entry, so there was no danger to the returning crew. Back on Earth the head of Roscosmos stated that he knew exactly what happened, but he was keeping it secret. Then a Russian tabloid published an article stating that a hole had been drilled by American astronaut Serena Auñón-Chancellor, which was immediately refuted by NASA who called it preposterous. As can be imagined, the atmosphere in the space station was quite tense. Rumours and theories were abound regarding the mystery hole in the spacecraft. Sergey Prokopyev was quick to point out that it was extremely difficult to drill a hole in microgravity and it was hard to imagine someone moving through the narrow confines of the space station carrying a drill. One theory was that the hole had been drilled during manufacture, either inadvertently or deliberately. If that was the case, it brought into question the scrutiny and safety of the manufacturing or in this case the lack of it. The crews however managed to put aside their thoughts and on 4 October 2018, the *Soyuz MS-08* crew of Oleg Artemyev, Andrew Feustal and Richard Arnold climbed into their spacecraft, undocked and returned to Earth. Three hours later their spacecraft touched down near Dzhezkazgan, Kazakhstan. One week later *Soyuz MS-10* was launched on 10 October 2018 and aborted minutes after lift-off. It was discovered that a booster rocket had failed, causing a collision at the point of separation between the second and third stages on the Soyuz-FG rocket. Seconds after the incident the capsule containing the two members of the crew, Aleksey Ovchinin (R) and Nick Hague (A), performed an emergency separation and returned to Earth in a ballistic trajectory, subjecting both crew members to between 6g and 7g. The crew were recovered safely and unharmed after landing near the town of Jezkazgan, Kazakhstan. After a lengthy investigation, during which time all missions to the ISS

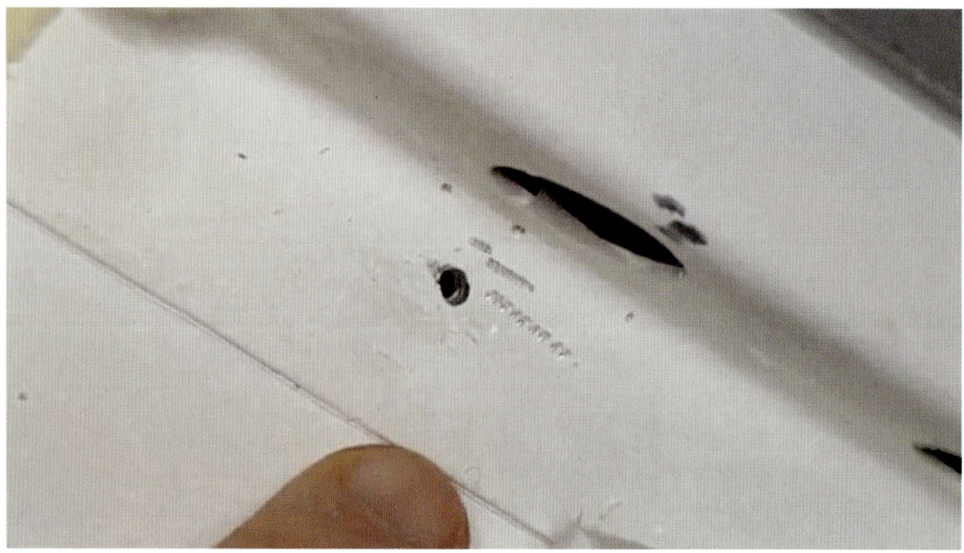

The drilled hole alleged to have been created by American astronaut Serena Auñón-Chancellor. (RKS Energia)

were temporarily suspended, it was concluded that the fault was due to the incorrect mating of the first stage booster to the first stage core. Other anomalies came to light such as a faulty cable connected to the booster and a deformed ball joint that had prevented proper separation. It was determined that as the cause was most likely down to an assembly error, no modification to the Soyuz-FG rockets were required. However concerns about the assembly of the rockets that carried both crews and supplies to the ISS became a worrying factor and caused some delays. Delivery of essential cargo to the ISS became a priority, although there were sufficient supplies on the space station to last for a couple of months. On 16 November *Progress MS-10* was launched aboard a Soyuz-2.1a rocket and, after a two day flight, docked in the aft port of the Zvezda module. Once secured the cargo spacecraft was unloaded of its cargo of water, fuel, oxygen and dry cargo, then slowly over the next few months, refilled with unwanted materials and waste. The Progress cargo spacecraft undocked on 4 June and was sent into a decaying orbit where it burnt up on re-entry.

A new date for the next manned mission was set and on 3 December 2018 when *Soyuz MS-11* was launched from Baikonur on a Soyuz-FG rocket with the *Expedition 58* of Oleg Kononenko (R), David Saint-Jacqes (Can) and Anne McClain (A). The mission had been scheduled for 20 December, but because of the accident to *Soyuz MS-10* had been brought forward. After a six-hour flight, the spacecraft docked in the zenith port of the Pirs module. With all the safety checks completed the relief crew cracked the hatch and entered the space station to be greeted by the resident crew. With the formalities over, both sets of crews got down to work on their experiments and mundane housekeeping tasks. On 20 December, the *Soyuz MS-09* crew, Sergei Prokopyev, Alexander Gerst and Serena Auñón-Chancellor, handed over command of the space station and prepared to leave. They climbed into their spacecraft and undocked; three hours later they landed safely near Dzhezkazgan, Kazakhstan. The problem of the 'drilled hole' was never resolved.

Progress MS-09, which had been docked with the ISS since 10 July 2018, was undocked on 29 January 2019 and placed into a decaying orbit. It entered the Earth's atmosphere later the same day and was burnt up on re-entry, the debris falling into the Pacific Ocean.

The next to visit the International Space Station was the *Expedition 59* crew of Aleksey Ovechinin (R), Christina Koch (A) and Nick Hague (A) on Soyuz MS-12. Launched on 14 March 2019 from Baikonur on a Soyuz-FG rocket, the spacecraft docked in the nadir port of the Rassvet module just six hours later. After the usual safety checks, the hatch was cracked for the relief crew to be greeted by the resident crew. This mission was initially scheduled to take the first astronaut from the United Arab Emirates (UAE), Hazza Al Mansoun or Sultan Al Neyadi as he was also known, but the accident concerning *Soyuz MS-10* had thrown listing into disarray and so his flight was put back to *Soyuz MS-15*. After the two crews had gone through all the formalities they settled down to carry out their experiments and tests, as well as keeping up with the maintenance and other housekeeping duties. The arrival of the cargo craft *Progress MS-11*, which was launched on 4 April 2019 on a Soyuz-2.1a rocket, brought 3,400 kilograms of fresh supplies and parts to the ISS. The unmanned spacecraft docked with the nadir port of the Pirs module just three hours and twenty-two minutes after leaving the launch pad at Baikonur. After being unloaded, it was refilled over the next few months with unwanted and discarded materials and waste.

Crew of *Soyuz MS-13*, L-R: Andrew Morgan, Aleksandr Skortsov and Luca Parmitano. (Roscosmos)

Soyuz MS-13 lifting off. (NASA)

On 24 June, the crew of Soyuz MS-11 handed over command of the space station, climbed into their spacecraft, undocked, and three hours later landed safely near Dzhezkazgan, Kazakhstan. It was almost a month later before the *Expedition 60* crew, Aleksandr Skvortsov (R), Luca Parmitano (It) and Andrew Morgan (A), arrived in *Soyuz MS-13*, which was launched on 20 July 2019 on a Soyuz-FG rocket, on the fiftieth anniversary of the First Moon Landing, After a six-hour rapid flight, the spacecraft docked in the aft port of the Zvezda module. With safety checks completed, the relief crew entered the ISS to be greeted by the resident crew. With formalities over, both crews settled down to their tasks. This had been intended to be the last purchased seat by NASA, but it was decided to purchase two more on upcoming flights. On 29 July *Progress MS-11* was undocked and placed in a decaying orbit to be burnt up on re-entry over the Pacific Ocean. Two days later on 31 July, *Progress MS-12* was launched on a Soyuz-2.1a rocket and scheduled for a three-hour flight to the space station. On arrival the unmanned cargo spacecraft docked in the Pirs module where, after safety checks, it was quickly unloaded. The cargo consisted of the usual food, water, oxygen and fuel, but also carried spare parts, medical equipment and personal items for the crew.

The next spacecraft to visit the ISS was an unmanned *Soyuz MS-14*, which was launched on a Soyuz-S.1a rocket on 22 August. It wasn't exactly unmanned as it had a robot called Feodor on board. After a two-day flight, a problem arose when *MS-14*'s first docking attempt to the Pirs module failed after a problem with the Kurs-P rendezvous system. The *Soyuz MS-13* crew then got into their spacecraft and undocked it from the aft port of the Zvezda module and then performed a manual docking in the Piris module. This then allowed *Soyuz MS-14* to carry out an automatic docking in the aft port of the Zvezda module. The mission of the unmanned *Soyuz MS-14* was to carry out the final test of the Soyuz-2.1a rocket and its abort system. It also tested an upgraded version of the navigation and propulsion control systems. The results were to be used in the design and development of a new Soyuz uncrewed and returnable cargo spacecraft, the Soyuz GVK. The remaining space was taken up with 1,450lbs of supplies for the space station. With the spacecraft safely docked and its cargo removed, the Robot Feodor remained inside. Aleksei Ovchinin carried out a series of tests using a special exoskeleton suit and a virtual-reality helmet connected to cameras on the robot and designed for the remote control of the robot in the so-called 'avatar' mode. On August 28, cosmonaut Aleksandr Skvortsov tried on the suit, which was plugged into a laptop computer. Initially there were problems activating the machine because the power button was stuck and at one point Aleksei Ovchinin proposed hitting the jammed button with a hammer. After over a dozen attempts he was finally able to release the button and activate the robot to function in a normal fashion. With all the test successfully completed the robot Feodor returned to Earth on 6 September in *Soyuz MS-14*, landing near Dzhezkazgan, Kazakhstan. Three weeks later the manned spacecraft *Soyuz MS-15* was launched carrying the *Expedition 61* crew of Oleg Skripochka (R), Hazza Al Mansoori (UAE) and Jessica Meir (A) to the ISS. The spacecraft was launched on a Soyuz-FG rocket from Baikonur on 25 September 2019 and, after a six hour rapid flight, docked in the aft port of the Zvezda Module. After the usual safety checks, the crew emerged from their spacecraft to be greeted by the resident crew. With the formalities over both crews settled down to their tasks and the mandatory housekeeping duties that all the crews had to adhere to. On 3 October the two members of the crew of *Soyuz MS-12*, Aleksei Ovchihih and Nick Hague, were joined by Hazza Al Mansoori who then undocked from the space station and three hours later landed safely near Dzhezkazgan, Kazakhstan.

Life aboard the space station continued as normal, then on 29 November the last of the waste and rubbish was dumped into *Progress MS-12*. The cargo spacecraft was then undocked and placed in a decaying orbit to burn up over the Pacific Ocean. This made way for the next arrival to the Intentional Space Station, the unmanned *Progress MS-13* on 6 December 2019. Launched on a Soyuz-2.1a rocket, the cargo spacecraft carrying 2,480 kilograms of cargo was originally scheduled for a fast delivery of three hours, but because of the expected arrival of the American spacecraft SpaceX *CRS19*, it was placed on a slow three-day flight and docked in the nadir port of the Pirs module on 9 December 2019.

On 6 February 2020, the crew of *Soyuz MS-13*, Alexsandr Skortsov, Luca Parmintano and Christina Koch, said their farewells to the now resident crew on the ISS, climbed into their spacecraft and three hours later landed back on Earth. Christina Koch had spent 328 days in space, breaking the previous record of 289 days held by American astronaut Peggy Whitson. Eleven days later, on 17 February, three more members of the International Space Station, Oleg Skripochka, Jessica Meir and

Soyuz MS-13 after landing. (NASA)

Andrew Morgan, said their farewells, climbed into the *Soyuz MS-15* spacecraft and undocked. Three hours later their spacecraft touched down safely near the town of Dzhezkazgan, Kazakhstan. The following month saw the arrival of *Expedition 62* with Anatoli Ivanishin (R), Ivan Vagner (R) and Christopher Cassidy (A) on board. Launched on a Soyuz-2.1a rocket on the 9 April 2020, *Soyuz MS-16* arrived at the space station just six hours later and docked in the Zenith port of the module. Because of the COVID pandemic that was sweeping across the world, restrictions were in place for the launch, which meant that no press or families were allowed to be present in the control centre to witness the event. With all safety checks completed, the crew entered the space station to the usual welcome given to the relief crew. With all the formalities completed the crews settled down to their work and the running of the space station. The arrival of *Progress MS-14*, which had been launched on 25 April 2020 on a Soyuz-2.1a rocket, docked in the aft port of the Zvezda module after a three hour and forty minute flight. After the cargo had been unloaded, the crew started to fill the cargo spacecraft with unwanted materials and waste. At the beginning of July, concerns were raised when debris from the break up of a Proton rocket, that had been launched in September 1987 to put three Glonass satellites in orbit, was found to be heading towards the space station. The objects were from forty-two pieces of debris that had been catalogued and tracked from the break up of the Proton rocket and were predicted to pass within one kilometre of the space station. The problem passed without incident to the relief of the crew on the International Space Station and to those on the ground. A program called COLA (Collision Avoidance) had always been in place, so on 3 July, *Progress MS-13* fired up its engine and raised the ISS orbit by one kilometre. On 8 July 2020, after the successful completion of the manoeuvre, *Progress MS-13* was undocked and placed in a decaying orbit to burn up over the Indian Ocean on re-entry. This then freed up the docking port in the Pirs module for the arrival of *Progress MS-15* which was launched on 23 July 2020 on a Soyuz-2.1a rocket. Following two orbits of the Earth, the cargo spacecraft docked in the nadir port of the Pirs module in just three hours and twenty minutes. After being unloaded it was left until 9 February 2021, during which time it was filled with waste and other unwanted materials, before being undocked and placed in a decaying orbit. It later re-entered the atmosphere and burnt up over the South Pacific Ocean.

It wasn't until 14 October 2020 that the next relief crew arrived. *Soyuz MS-17* was launched on a Soyuz-2.1a rocket to deliver the *Expedition 63* crew of Sergey Rhyzikov (R), Sergey Kurd-Skverchkov (R) and Kathleen Rubins (A) to the International Space Station. This flight was the first to use the 'ultra-fast' flight plan which was designed to take the spacecraft to the space station in just three hours. The Soyuz manned spacecraft successfully docked in the nadir port of the Rassvet module right on schedule. After completing the usual safety checks, the crew entered the space station to be greeted by the resident crew. With the formalities over the crews settled down to their tasks as well as the maintenance duties required by all members of the crew. Almost nothing is known about the experiments that were undertaken, save for cardiovascular and physiological tests that were regularly carried out. On 22 October, with all their experiments and research completed, the crew of *Soyuz MS-16* handed over command of the space station, said their farewells, climbed into their spacecraft and returned to Earth. They touched down on the steppes of Kazakhstan just over three hours later.

It was to be four months before the next arrival and that was the cargo carrying spacecraft *Progress MS-16*, which had been launched on 15 February 2021 on a Soyuz-2.1a rocket. This was not going to be one of the rapid flights and it wasn't until 17 February that the cargo spacecraft finally docked in the nadir port of the Pirs module. The crew unloaded the 2,460 kilograms of cargo, which consisted of fuel, drinking water, the usual food and personal items and a repair kit with reinforced patches for temporary sealing of the transfer chamber in the Zvezda module. It was to remain docked until 26 July, when it was undocked, taking the now decommissioned Pirs module with it. Both the module and the cargo spacecraft deorbited and were burnt up on re-entry the same day. This was the first time a module had been decommissioned and it was to be replaced with the Nauka module. In the meantime the intended arrival of *Soyuz MS-18* meant that the *Soyuz MS-17* spacecraft had to be moved from the Rassvet module to the Pirs module, which was carried out on 19 March 2021. Three weeks later, on 9 April 2017, *Soyuz MS-18* lifted off the launch pad on a Soyuz-2.1a rocket carrying the *Expedition 64* crew members Oleg Novitsky (R), Pyotr Dubrov (R) and Mark Vande Hei (A) to the ISS where it docked in the nadir port of the Rassvet module. After the usual safety procedures had been completed, the relief crew entered the space station. With all the greetings and formalities completed the crews got down to work. For the *Soyuz MS-17* crew it was a case of winding down their experiments and research work, as nine days later they handed command of the space station to the new crew and prepared to return to Earth. On 17 April 2021 *Soyuz MS-17* undocked from the space station and three hours later touched down safely on the Kazakh Steppe, Kazakhstan.

It wasn't until 29 June 2021 that the next visitor arrived at the International Space Station in the shape of *Progress MS-17*. Launched on a Soyuz-2.1a rocket, the cargo spacecraft took just three hours to reach the space station where it docked in the zenith port of the Poisk module. Amongst the usual supplies were a number of pieces of film equipment in readiness for a short film to be made aboard the ISS. After being unloaded, the cargo spacecraft remained docked until the 25 November, when, after being refilled with waste materials, it undocked to leave the port free for the arrival of *Progress M-UM*. Three weeks after the arrival of *Progress MS-17* on 21 July, a new section of the ISS arrived aboard a Proton-M rocket: the Nauka module, also known as the Multipurpose Laboratory Module-Upgrade (MLM-U). The new module docked with the ISS on 28 July along with the European Robotic Arm (ERA) in the nadir port of the Zvezda module. This was the first expansion of the Russian segment of the ISS for over twenty years. A problem arose just after Nauka docked when it started firing its engine thrusters, causing the space station to make a one and a half rotation before the thrusters ran out of fuel. Ground controllers managed to stop the rotation, enabling the ISS crew to regain control. The module had two docking ports, an active, which was used to attach the module to the station using the nadir port of the Zvezda module, and a passive one which was used to attach the Prichal module which was delivered by *Progress MU-M* in the November. *Progress M-17* was to remain docked until 21 October when it was moved to another of the Russian segments of the space station, the Nauka module. In the meantime *Soyuz MS-19* had been launched on a Soyuz-2,1a rocket on 5 October 2021. This was to be one of the most controversial missions to the International Space Station. On board were Anton Shkaplerov (R), Kim Shipenko – film director (R) and Yulia Peresild – actress (R). Talks with Roscosmos had been going on for some months regarding the making of a film showing that ordinary people could be trained in three or four months to be able to work in space. The proposal received opposition from both the scientific and aerospace companies, because they said that not only did it remove trained cosmonauts from their flights, it was a misuse of public money and they were using the space station's resources not for the benefit of mankind but for non-scientific purposes, which in their eyes was illegal. Sergey Krikalev, who was director of crew operations at Roscosmos at the time, was particularly vociferous in his objections and was said to have been dismissed. A few days later he was reinstated after some particularly equally vociferous protests from engineers and cosmonauts who were either on or off active duty. The *Soyuz MS-19* space craft docked after a three hour flight, in the nadir port of the Rassvet module. After the usual safety checks, the oncoming crew cracked the hatch and entered the space station to be welcomed by the resident crew. After all the formalities were completed everything settled down to normal – well, as normal as could be with a film being made. Because of the relatively cramped conditions on the space station, care had to be taken whilst they were filming not to get in the way of the scientific experiments that were still being conducted. Kim Shipenko, who was director, camera operator, art director and make-up artist, shot about 35–40 minutes of the film called *The Challenge*. The remainder of the crew became 'extras' in some scenes. On 17 October, after completing their filming, Oleg Novitsky and the two members of the film crew, Kim Shipenko and Yulia Peresild, said their farewells, climbed into the *Soyuz MS-18* spacecraft, undocked and returned to Earth. The spacecraft landed safely three hours later on the Kazkha Steppe, Kazakhstan. It was discovered later that the director of Channel One, Dimitry Rogozin, had paid for Kim Shipenko's and Yulia Peresild's seats.

The arrival of *Progress MS-18*, launched on a Soyuz 2-1a rocket on 28 October 2021, was a welcome sight to the resident crew of the ISS. After a two-day flight, the unmanned cargo spacecraft docked in the aft port of the Zvezda module. It had been planned for a three-hour flight, but an air leak in the docking module was still the cause for some concern, causing it to be delayed. After being unloaded of fresh food, water, fuel and personal items, the cargo spacecraft remained docked until 1 June 2022, all the while being used as a receptacle for unwanted materials and rubbish. It also delivered a part of a work platform for the MLM (Multi-purpose Laboratory Module). Another part of the platform was later delivered on *Progress MS-21*. The whole unit was later transferred over to the Nauka module where it was re-installed in the base of the European Robotic Arm (ERA) where it was used to mount payloads on the outside of the Nauka module. After undocking on 20 October 2021, *Progress MS-17* went into a decaying orbit and was burnt up as it entered the Earth's atmosphere on 25 November. *Progress MS-18* remained until 1 June 2022, when it was undocked and placed in a decaying orbit before entering the Earth's atmosphere and being destroyed over the South Pacific Ocean.

Just under one month later, on 24 November, another Progress cargo spacecraft was launched, *Progress M-UM*. Launched on a Soyuz-2.1a rocket, this cargo spacecraft was carrying the Prichal module that was to be an extension of the Russian Orbital Segment of the ISS. As the spacecraft was launched, *Progress MS-17* was undocked from the nadir port on the Nauka module, taking with it the special extension from the Nauk's docking mechanism that had been designed specifically for cargo-carrying and manned spacecraft. This meant that the Prichal module could not dock in the SSVP (Sistema Stykovki i Vnutrennego Perekhoda) port, the Russian segment of the ISS, in the Rassvet module. After some configuration, the Prichal module was able to dock in the Nanuka module. On 25 November *Progress MS-17* was put into a decaying orbit and left to burn up over the Pacific Ocean.

Two more paying space tourists arrived on the International Space Station on 8 December 2021. Launched from Baikonur on a Soyuz-2.1a rocket, *Soyuz MS-20* with Alexsnder Misurkin (R), Yusaku Maezawa (Jpn) and Hizano Hirano (Jpn) on board, the spacecraft docked with the ISS in the zenith port on the Poisk module after a six-hour flight. This was the first Russian space mission dedicated purely to space tourism. Up to this point, one space tourist would be taken and would return after a short stay as part of a returning crew. After eleven days the crew of *Soyuz MS-20* undocked and three hours later landed safely on the Kazakha Steppe, Kazakhstan. There must have been periods when the tourists got in the way of the resident crew members, who were not only conducting experiments, but trying to maintain the space station. I have no doubt that the tourists were made to play their part in the housekeeping duties that were necessary for the benefit of all the occupants of the space station. However it has to be remembered that the tourists brought much needed funds to the Russian Space Agency, so it was a cross they had to bear. Space tourists were not the only ones paying for seats on the Soyuz spacecraft: NASA and the ESA were also having to fund their astronauts, scientists and engineers.

It wasn't until 15 February 2022 that the next spacecraft arrived at the space station and that was the cargo spacecraft *Progress MS-19*, which was launched on a Soyuz-2.1a rocket. This was one of the fast tracked missions and the cargo spacecraft docked in the zenith port of the Poisk module just three hours and twenty minutes after launch. Along with the usual cargo of food, water, oxygen, fuel experimental and personal items, were six Russian experimental 'cubesats', that were deployed by Oleg Artemyev during his third spacewalk. The 'Cubesat' was a small communication satellite of just 39in (10cm), weighing 4.4lb (2kg) and constructed using commercially available off-the-shelf components.

The next spacecraft to arrive at the ISS was the manned *Soyuz MS-21*, which was launched on 22 March 2022 from Baikonur on a Soyuz-2.1a rocket. The *Expedition 66* crew were all Russians: Oleg Artemyev, Denis Matveev and Sergey Korsakov. Originally NASA were going to purchase a seat on the spacecraft for American astronaut Loral O'Hara who would have replaced Sergey Korsakov, but they decided to turn down the offer. The situation in the Ukraine may have had some bearing on the American's decision. Three hours and twenty minutes after launching, the spacecraft docked in the nadir port of the Prichal module. After going through all the safety procedures, the relief crew cracked the hatch and entered the space station to be greeted by the resident crew. There was some controversy over the colours of the relief crew's flight suits which were normally blue, these however were bright yellow with blue segments. A number of international commentators who were covering the launch said that these were the colours of the Ukrainian flag and wondered if the three cosmonauts were showing their support for the Ukrainian people in their fight against the invading Russian Army. There was already a very tense atmosphere concerning space co-operation for the International Space Station after international sanctions had been imposed on Russia. Roscosmos was quick to point out that they were the identifying colours of the Bauman Moscow State Technical University from which all the three cosmonauts had graduated. With all the formalities over, the crews settled down to the normal day-to-day running of the space station. At the end of March, two members of the *Soyuz MS-18* crew, Mark Vande Hei and Pytor Dubrov, were joined by Anton Skhapierov from *Soyuz MS-19*, and, after saying their farewells,

climbed into their spacecraft and undocked from the ISS. After a three-hour flight they landed safely in Kazakhstan.

It was to be a further two months before the next spacecraft arrived at the ISS. It was the unmanned *Progress MS-20* cargo spacecraft which was launched on 3 June 2022 on a Soyuz-2.1a rocket. Three hours and twenty minutes later the cargo spacecraft, carrying 2,500 kilograms of cargo, automatically docked in the nadir port of the Zvezda module. Amongst the usual water, oxygen, fuel and food, was a 3D printer and four more experimental 'cubesats', which were later deployed during a spacewalk by Oleg Artemyev. To date over 3,500 of these small satellites have been launched, making the orbit around the Earth a veritable traffic jam of spent and active satellites together with remnants of other pieces of redundant and discarded space material. After being unloaded, the cargo spacecraft was left docked and gradually filled with waste and other unwanted materials. This meant that there were now two cargo spacecraft docked with the ISS, so *Progress MS-19*, which had been left docked since 15 February, during which time it too had been filled with the usual waste materials and unwanted items, was undocked on the 23 October and within hours had gone into a decaying orbit to be burnt up over the South Pacific Ocean. In the meantime *Soyuz MS-22* was launched on 22 September 2022 on a Soyuz-2.1a carrier rocket. The spacecraft carried the next relief crew of Sergey Prokopyev (R), Dimitri Petelin (R) and Francisco Rubio (A). After a three hour and eleven minute flight, *Soyuz MS-22* docked in the nadir port of the Rassvet module. With the usual safety procedures completed, the relief crew entered the space station; this was Sergey Prokopyev's second visit so was well accustomed to the situation there. With all the formalities over, the crews settled down to work and carry out the housekeeping duties that all crew members were faced with when on the ISS. One month later the cargo spacecraft *Progress MS-21* was launched on 26 October 2022 on a Soyuz-2.1a rocket. This was not one of the rapid launches, but one that took two days before automatically docking in the zenith port of the Poisk module. As with all previous cargo spacecraft, after it was unloaded it remained docked as a refuse container. Then on 15 December a problem arose when a visible stream of what were described as 'flakes' was seen emanating from the *Soyuz MS-22* spacecraft. At the same time it was noticed that there was a loss of pressure in the external radiator cooling loop. Using the cameras mounted on the Canadarm, the analysis of the images allowed engineers to detect a possible section of damage on the surface of the service module. It was discovered later that the leak had been caused by a micro-meteorite strike leaving a 0.8mm (0.031in) hole in the external radiator on the service module. This incident showed how fraught with danger was living in a space environment and, with more and more debris and other objects being dumped in space, the risk was intensified. *Soyuz MS-22* was deemed to be unsafe to carry any crew, so it was decided to return it to Earth unmanned. Initially there appeared to be a problem in getting the crew off the space station, but during this period, the Americans had docked with the ISS in their *SpaceX Dragon* spacecraft which had the capability to carry seven crew members. In the event that there was a problem, astronaut Francisco Rubio's seat liner was removed from the *Soyuz MS-22* and placed in SpaceX. In the meantime *Progress MS-22*, which had been launched on 9 February 2023 on a Soyuz MS-2.1a rocket, arrived. This was not a rapid flight, but a two-day flight and the cargo carrying spacecraft automatically docked in the aft port of the Zvezda module. After being unloaded it was left docked to be used as a large expensive waste bin. With the problem on *Soyuz MS-22* still to be resolved, the *Soyuz MS-23* manned mission was put on hold and the unmanned spacecraft sent to the ISS to act as a lifeboat. Launched on the 24 February 2023 on a Soyuz-2.1a rocket, the spacecraft docked in the zenith port of the Poisk module on the 25 February. It was relocated to the nadir port on the Prichal module on 6 March. With the arrival of *Soyuz MS-23* at the Prichal module, Francisco Rubio's seat liner was transferred from *SpaceX Dragon* and placed in *Soyuz MS-23*. It has to be remembered that each seat liner had been moulded to fit the contours of the cosmonaut's body so it was necessary to move it. The unmanned damaged *Soyuz MS-22* was undocked from the ISS on 6 April 2023 and returned to Earth, landing on the Kazakh Steppe, Kazakhstan. Although the spacecraft was unmanned, it was used to transport a large number of experimental packages that had been completed, together with a number of other items. Whilst the drama on the ISS was unfolding, the next cargo spacecraft arrived, *Progress MS-23*. Launched on 24 May 2023 on a Soyuz-2.1a rocket and after a three hour twenty minute flight, the cargo spacecraft docked in the zenith port of Poisk module. It carried the usual cargo of food, fuel, oxygen and personal equipment and after being unloaded remained docked to be used as a waste container.

On 10 August 2023 attention moved away from the International Space Station, when an unmanned Soyuz 2.1b rocket was launched from the Vostochny Cosmodrome in the Amur region of East Russia, carrying *Luna 25* to the Moon. The last lunar mission was *Luna 24* which was launched almost fifty years ago and had successfully returned to Earth with more samples of lunar material. *Luna 25* was scheduled to stay on the Moon's surface for at least a year, during which time it would collect and analyse samples and send the results back to Earth. The cameras that were installed on the spacecraft transmitted a number of images of both the lunar surface and of the Earth whilst in lunar orbit. Unfortunately, whilst making its descent onto the Moon, the spacecraft suffered a problem with the descent engine and with its computer guidance system, causing it to crash onto the lunar surface. All communication with the spacecraft was lost and it was deemed to have been destroyed.

Work on the ISS, however, continued and three months after the arrival of *Progress MS-23*, another cargo craft arrived, *Progress MS-24*, which had been launched on 23 August 2023 on a Soyuz-2.1a rocket. After a two-day flight, the spacecraft docked in the aft port of the Zvezda module. The cargo craft was unloaded and it too was left docked to the space station to be used as a waste dump. It wasn't until 23 February 2024 before it was undocked and immediately placed in a decaying orbit to be burnt up the same day over the Pacific Ocean.

The next crew, Oleg Kononenko (R), Nikolai Chub (R) and Loral O'Hara (A), arrived on 15 September 2023 in *Soyuz MS-24*, which was launched on a Soyuz-2.1a rocket. A gift of three small plush-covered model seagulls were given to the crew prior to lift-off, by the Director of the Chekhov Moscow Art Theatre, and served to indicate weightlessness once the spacecraft had left Earth. Three hours after lift-off, the Soyuz spacecraft docked in the nadir port of the Rassvet module. After the usual safety checks, the crew opened the hatch to be greeted by the resident crew and the crew of *SpaceX-7 Dragon*, who had arrived earlier. Once in the space station, Kononenko, Chub and O'Hara became part of the *Expedition 69* crew of Sergey Prokopyev (R), Dimitri Petelin (R) and Francisco Rubio (A) and with the formalities over, settled down to their tasks and experiments in a now crowded ISS. Twelve days later on 27 September 2023, the crew of *Soyuz MS-22*, Sergey Prokopyev, Dimitri Petelin and Francisco Rubio, said their farewells to the remaining ISS crew members, climbed into the *Soyuz MS-23* spacecraft and undocked. Three hours later the spacecraft safely touched down on the Kazakh Steppe, Kazakhstan. *Progress MS-23* was released from the ISS on 29 November 2023 and placed in a decaying orbit to burn up the same day over the Pacific Ocean. On 12 March 2024 *SpaceX-7* undocked from the ISS and returned to Earth with Jasmin Moghbeli (A), Andreas Mogensen (ESA), Satoshi Furukawa (Jpn), and Konstantin Borisov (R) and splashed down safely in the Gulf of Mexico off the coast of Pensacola, Florida, after completing a six-month science mission.

The next visit of a manned spacecraft, *Soyuz MS-25*, had been scheduled for 21 March 2024, but a voltage drop in one of the power generators caused the launch to be abandoned. It was re-scheduled for launch two days later and was successful. It carried a crew consisting of Oleg Novitsky (R), Tracy Caldwell-Dyson (A) and space tourist Maryna Vasileuskaya of Belarus. Vasileuskaya, a former flight attendant, was selected to represent the Republic of Belarus, under an agreement with Roscosmos. After spending a week on the station, Marina Vasileyeva returned to Earth in *Soyuz MS-24* with Loral O'Hara and Oleg Novitsky. Their spacecraft touched down near the remote town of Dzhezkazgan, Kazakhstan. NASA astronaut Tracy Caldwell-Dyson was scheduled to return to Earth with Oleg Kononenko and Nikolai Chub on *Soyuz MS-25* on 25 September 2024. Those remaining on the ISS are NASA astronauts Michael Barratt (*SpaceX-8*), Matthew Dominick (*SpaceX-8*), Tracy Dyson and Jeannette Epps (*SpaceX-8*) as well as Russian cosmonauts Nikolai Chub, Alexander Grebenkin and Oleg Kononenko.

The war in the Ukraine is now having a detrimental effect on the Russian Space program, as a number of countries have said that they will refuse to co-operate with Russia whilst the conflict continues. This of course creates a major problem with the International Space Station as Russia is the prime mover of crews and supplies. Where the Soviet Space program goes from here is anyone's guess, but hopefully common sense will prevail and the knowledge acquired regarding space exploration will not be wasted. At the beginning of 2024, a NASA spokesman announced:

'NASA and Roscosmos have amended the integrated crew agreement to allow for a second set of integrated crew missions in 2024 and one set of integrated crew missions in 2025, for continued safe operations of the space station, the integrated crew agreement helps ensure that each crewed spacecraft docked to the station includes an integrated crew with trained crew members in both the Russian and US Operating Segment systems.'

Hopefully this agreement will be extended even further, but only time will tell.

Bibliography

Evgeny Riabechkov. *Russians in Space.* The Novosti Press, Moscow. ISBN 0 297 99375.

In Gagarin's Trail. Novosti Press, Moscow.

Krieger, F.J. *Casebook on Soviet Aeronautics Part 1 & 2,* Santa Monica, Rand Corporation, USA, 1956.

Solkolsky, V.N. *Short Outline of the Development of Rocket Research in the USSR*. Moscow Academy of Science 1960.
Translated by the U.S. Department of Commerce.

Solkolsky, V.N. *Russian Solid-Fuel Rockets.* Moscow Academy of Science of the USSR 1964.

Stoiko, Michael. *Soviet Rocketry*. David & Charles, London, 1971.

Tsander, F.A. *Problems of Flight by Jet Propulsion.* Moscow Publishing House. 1961.

Tsiolkovsky, K.E. *Work on Rocket Technology.* Moscow Publishing House. 1947.

V.P. Glushko. *Rocket Engines GDL-OKR.* Academy of Science, Moscow.

Yuri Maximov. *Man and Space.* The Novosti Press, Moscow.

Index

R

S

T

U

V